DAVID HILBERT

FOUNDATIONS OF GEOMETRY

(Grundlagen der Geometrie)

SECOND ENGLISH EDITION

Translated
by
Leo Unger
from the Tenth
German Edition

Revised and Enlarged
by
Dr. Paul Bernays
Professor Emeritus
Eidgenössische Hochschule Zürich

OPEN COURT
LA SALLE • ILLINOIS

Originally published as *Grundlagen der Geometrie*, B.G. Teubner, Stuttgart. This is the sole authorized English translation. Translated from the tenth German edition, with permission of B.G. Teubner, Stuttgart.

Second printing 1980
Third printing 1987
Fourth printing 1988
Fifth printing 1990
Sixth printing 1992

Printed and bound in the United States of America.

Foundations of Geometry
Translated from the German by Leo Unger/Revised by Paul Bernays

Library of Congress Catalog Card Number 73-110344

ISBN: 0-87548-164-7

The present new printing of Hilbert's *Foundations of Geometry* is really not a new revision. Rather, only a few corrections and minor additions have been introduced.[1]

Some of the dependences which were stated before in Appendix VI within the set of axioms of the real numbers are grouped in Supplement I; Supplement III is essentially a reproduction of a corollary in Appendix II of the Fifth Edition concerning the deducibility of the axiom of congruence in the broader sense from that in the narrower sense by assuming the axioms of inclusion.

Supplement II has been added as a new and simplified form of the theory of proportion whose previous form has been retained in Sections 14-16 of the main text.

Of Appendices I-X of the Seventh Edition only those of geometric nature, I-V, have been retained.

In order to convey some idea about the Seventh Edition, the following is reprinted from Hilbert's own preface. "The present Seventh Edition of my book *Foundations of Geometry* brings considerable improvements and additions over the previous editions, partly from my subsequent lectures on this subject and partly from improvements made in the interim by other authors. The main text of the book has been revised accordingly. In this work one of my students, H. Arnold Schmidt, has assisted me most energetically. He has not only worked out the details for me, but he has also contributed numerous remarks and corollaries of his own, in particular, the new form of Appendix II has been formulated by him. I hereby express to him my sincerest thanks for his assistance."

Reference is also made to the historical survey given by A. Schmidt in *Hilbert's Collected Papers* (Berlin, 1933), Volume II, "Zu Hilberts Grundlegung der Geometrie," pp. 404-414.

Zürich, May 1956 P. Bernays

[1] In particular, the elimination of the order axioms from the affine development of the segment arithmetic (especially in Sections 24-26), which is theoretically possible, has been avoided.

Except for minor changes and additions to the text this Ninth Edition of Hilbert's *Foundations of Geometry* differs from the preceding one only in the supplements. The additions incorporated in these were motivated largely by the investigations of H. Freudenthal in his article "Zur Geschichte der Grundlagen der Geometrie," *Nieuw Archief voor Wiskunde* (4), V, 105-142 (1957), which he dedicated to the Eighth Edition of Hilbert's book, in particular by his criticism of the exposition of the theory of surfaces and its subsequent applications given there.

Elaborations on Sections 3 and 4 of the main text regarding the consequences of the axioms of incidence and order have been added in Supplement I. In particular, an observation is quoted there from the article "De logische Grondslagen der Euklidische Meetkunde" by Van der Waerden.

A remark about the arithmetic of ratios has been added at the end of Supplement II.

Supplement III introduces additional considerations to the theory of area.

Supplement IV, 1 examines the possibility of eliminating the axioms of order from the investigations in Chapter V and Supplement IV, 2 considers a remark by Kijne to Theorem 65 on construction problems.

Supplement V, 1 contains additional remarks concerning the two "non-Pythagorean" geometries constructed by Hilbert in Appendix II.

Supplement V, 2 is former Supplement III, with, however, a correction in a proof.

Some references to recent literature have been added.

Zürich, Spring 1962 P. Bernays

The last edition of Hilbert's *Foundations of Geometry* that appeared in his lifetime was the Seventh Edition. In order to convey some idea about it the following is reproduced from Hilbert's own Preface. "The present Seventh Edition of my book *Foundations of Geometry* brings considerable improvements and additions to the previous edition, partly from my subsequent lectures on this subject and partly from improvements made in the meantime by other writers. The main text of the book has been revised accordingly. In this regard, one of my students, H. Arnold Schmidt, has assisted me most energetically. He has not only done this work for me by himself but has also contributed numerous remarks and corollaries of his own; in particular, the new form of Appendix II has been independently written by him. I hereby express to him my sincerest thanks for his assistance."

Reference is also made to the historical survey, "Zu Hilberts Grundlegung der Geometrie" by A. Schmidt in *Hilbert's Collected Papers*, Volume II, pages 404-414, Berlin, 1933.

The Eighth, Ninth and Tenth Editions introduce no essentially new revisions. Only a few corrections and minor additions have been appended in them to the original text. However, a few supplements have been added. Of Appendices I-X of the Seventh Edition only those of geometric nature, I-V, have been retained.

The additions incorporated in the supplements were largely inspired by the investigations of H. Freudenthal in his article "Zur Geschichte der Grundlagen der Geometrie," *Nieuw Archief voor Wiskunde* (4), pages 105-142 (1957), which he dedicated to the Eighth Edition of Hilbert's book, in particular by his critique of the exposition of the theory of surface area and its subsequent applications given there.

As far as the content of the supplements is concerned, enlargements of Sections 3 and 4 of the main text, dealing with the consequences of the axioms of incidence and order, have been added in Supplement I, 1. In particular, a remark is taken there from Van der Waerden's article "De logische Grondslagen der Euklidische Meetkunde" (quoted there). Some dependences among the axioms of the real numbers, previously introduced in Appendix VI, have been put in Supplement I, 2.

Supplement II brings a new simplified form of the theory of proportion whose previous form has been retained in Sections 14-16 of the main text.

Supplement III contains additional considerations about the theory of surface area.

Supplement IV, 1 examines the possibility of eliminating the axioms of order from the investigations of Chapter V and Supplement IV, 2 is a sharpening of Theorem 65 (in Section 37) on construction problems based on a remark by D. Kijne.

Supplement V, 1 contains additional remarks on the two non-Pythagorean geometries constructed by Hilbert in Appendix II.

Supplement V, 2 is essentially a reproduction of an addendum to Appendix II of the Fifth Edition. It deals with the possibility of deducing the broader congruence axiom from the narrower one by assuming the axiom of embedment and at the same time makes a correction to a proof.

Some references to recent literature have been added.

Zürich, February 1968 P. Bernays

CONTENTS

FOREWORD

All of the mathematical disciplines are historically determined. Geometry especially has had a long and turbulent development, from the time of Euclid and the early commentators, through the medieval universities into modern times. The work of Hilbert certainly was of major importance in altering man's concept of Geometry as, in a sense, idealistic truth. Starting with the first edition of his Foundations of Geometry, Hilbert's book itself has undergone major changes, so that this latest edition is hardly recognizable as the first edition. For this reason it is of great importance for the serious college student and high school teacher to have available in English this latest edition.

Many mathematicians have expressed the opinion that Hilbert's work is elementary or of small importance, full of error, and devoid of modern significance. With all respect to the men who have made such strictures, I should like to emphasize the great importance of an attempt at developing a complete consistent statement of the axioms of geometry and a synthesis of these axioms within the analysis of real number. As all of us have often wondered when first faced with analytic geometry after high school synthetic geometry, I puzzled for years about the connection between the two. Hilbert went further than puzzling. Much earlier than my puzzling, he proved that in the analysis of real number and, in particular, in the algebra of three real variables there is a possible model for the axioms of geometry as he revised them. More than this, he showed how to establish that this model is truly unique, that is, any model is isomorphic to it.

There is confusion in the language Hilbert used in his first edition and some of this confusion exists up through this one. Not only must the terms point, line and plane and the relation of betweenness be primitive; but also the relation of incidence said of lines and pairs of distinct points, must be primitive and yet distinct from the primitive relation of incidence said of planes and triples of noncollinear points; and the relation of congruence said of line segments must be primitive yet distinguished from the primitive relation of congruence said of angles.

This kind of thing is important; yet it does not detract too much from the usefulness of the text for the student who is forewarned that sufficient to the day is the rigor thereof. Hilbert was a giant and we are fortunate to be able to live in his shadow.

Professor of Mathematics
Oregon State University

Harry Goheen

All human knowledge thus begins with intuitions, proceeds thence to concepts and ends with ideas.

Kant, *Critique of Pure Reason,* "Elements of Transcendentalism," *Second Part,* II.

INTRODUCTION

Geometry, like arithmetic, requires only a few and simple principles for its logical development. These principles are called the **axioms** of geometry. The establishment of the axioms of geometry and the investigation of their relationships is a problem which has been treated in many excellent works of the mathematical literature since the time of Euclid. This problem is equivalent to the logical analysis of our perception of space.

This present investigation is a new attempt to establish for geometry a **complete**, and **as simple as possible**, set of axioms and to deduce from them the most important geometric theorems in such a way that the meaning of the various groups of axioms, as well as the significance of the conclusions that can be drawn from the individual axioms, come to light.

CHAPTER I

THE FIVE GROUPS OF AXIOMS

§ 1. The Elements of Geometry and the Five Groups of Axioms

DEFINITION. Consider three distinct sets of objects. Let the objects of the **first** set be called *points* and be denoted by $A, B, C, \ldots$; let the objects of the **second** set be called *lines* and be denoted by $a, b, c, \ldots$; let the objects of the **third** set be called *planes* and be denoted by $\alpha, \beta, \gamma. \ldots$ The points are also called the *elements of line geometry*; the points and the lines are called the *elements of plane geometry*; and the points, the lines and the planes are called the *elements of space geometry* or the *elements of space*.

The points, lines and planes are considered to have certain mutual relations and these relations are denoted by words like **"lie," "between," "congruent."** The precise and mathematically complete description of these relations follows from the **axioms of geometry**.

The axioms of geometry can be divided into five groups. Each of these groups expresses certain related facts basic to our intuition. These groups of axioms will be named as follows:

I,	1 - 8	Axioms of *Incidence,*
II,	1 - 4	Axioms of *Order,*
III,	1 - 5	Axioms of *Congruence,*
IV,		Axiom of *Parallels,*
V,	1 - 2	Axioms of *Continuity.*

§ 2. Axiom Group I: Axioms of Incidence

The axioms of this group establish an *incidence* relation among the above-introduced objects–points, lines and planes, and read as follows:

I, 1. *For every two points A, B there exists a line a that contains each of the points A, B.*

I, 2. *For every two points A, B there exits no more than one line that contains each of the points A, B.*

Here as well as in what follows, two, three, . . . points or lines, planes are always to be understood as distinct points or lines, planes.

Instead of **"contains"** other expressions will also be used, e.g., a **passes through** A **and through** B, a **joins** A **and** B **or joins** A **with** B, A

lies on a, A **is a point of** a, **there exists a point** A **on** a, etc. If A lies on the line a as well as on another line b the expressions used will be '**The lines** a **and** b **intersect at** A, **have the point** A **in common**', etc.

I, 3. *There exist at least two points on a line. There exist at least three points that do not lie on a line.*

I, 4. *For any three points A, B, C that do not lie on the same line there exits a plane* α *that contains each of the points A, B, C. For every plane there exists a point which it contains.*

The expressions 'A **lies in** α; A **is a point of** α,' etc. will also be used.

I, 5. *For any three points A, B, C that do not lie on one and the same line there exists no more than one plane that contains each of the three points A, B, C.*

I, 6. *If two points A, B of a line a lie in a plane* α *then every point of a lies in the plane* α.

In this case it is said that **the line** a **lies in the plane** α, etc.

I, 7. *If two planes* α, β *have a point A in common then they have at least one more point B in common.*

I, 8. *There exist at least four points which do not lie in a plane.*

Axiom I, 7 expresses the fact that space has no more than three dimensions, whereas Axiom I, 8 expresses the fact that space has no less than three dimensions.

Axioms I, 1-3 can be called the *plane axioms of group* I in distinction to Axioms I, 4-8 which will be called the *space axioms of group* I.

Of the theorems that ensue from Axioms I, 1-8 only the following two are mentioned:

THEOREM 1. Two lines in a plane either have one point in common or none at all. Two planes have no point in common, or have one line and otherwise no other point in common. A plane and a line that does not lie in it either have one point in common or none at all.

THEOREM 2. Through a line and a point that does not lie on it, as well as through two distinct lines with one point in common, there always exists one and only one plane.

§ 3. Axiom Group II: Axioms of Order[1]

[1] These axioms were first studied in detail by M. Pasch in his *Vorlesungen über neuere Geometrie* (Leipzig, 1882). In particular, Axiom II, 4 is essentially due to him.

The axioms of this group define the concept of "**between**" and by means of this concept the *ordering* of points on a line, in a plane, and in space is made possible.

DEFINITION. The points of a line stand in a certain relation to each other and for its description the word *"between"* will be specifically used.

II, 1. *If a point B lies between a point A and a point C then the points A, B, C*

A B C

are three distinct points of a line, and B then also lies between C and A.

II, 2. *For two points A and C, there always exists at least one point B on the line AC such that C lies between A and B.*

A C B

II, 3. *Of any three points on a line there exists no more than one that lies between the other two.*

Besides these *line axioms of order* a *plane axiom of order* is still needed.

DEFINITION. Consider two points, *A* and *B*, on a line *a*. The set of the two points *A* and *B* is called a *segment*, and will be denoted by *AB* or by *BA*. The points between *A* and *B* are called the points of the segment *AB*, or are also said to lie *inside* the segment *AB*. The points *A, B* are called the *end points* of the segment *AB*. All other points of the line *a* are said to lie *outside* the segment *AB*.

II, 4. Let *A, B, C* be three points that do not lie on a line and let *a* be a line in the plane *ABC* which does not meet any of the points *A, B, C.* If the line *a* passes through a point of the segment *AB*, it also passes through a point of the segment *AC*, or through a point of the segment *BC*.

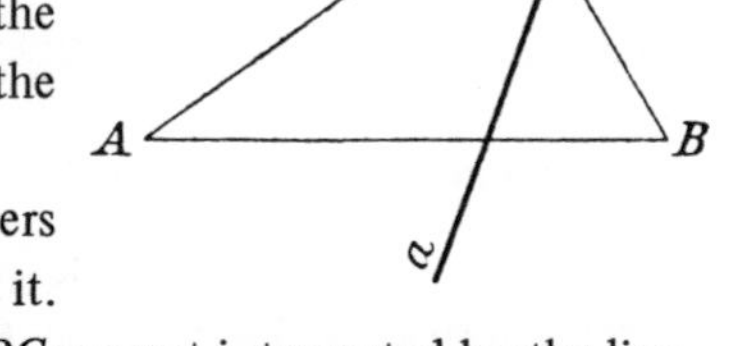

Expressed intuitively, if a line enters the interior of a triangle, it also leaves it. The fact that both segments *AC* and *BC* are not intersected by the line *a* can be proved. (See Supplement I, 1.)

§ 4. Consequences of the Axioms of Incidence and Order

The subsequent theorems follow from Axioms I and II:

THEOREM 3. For two points A and C there always exists at least one point D on the line AC that lies between A and C.

PROOF. By Axiom I, 3 there exists a point E outside the line AC, and by Axiom II, 2 there exists on AE a point F such that E is a point of the segment AF. By the same axiom and by Axiom II, 3 there exists on FC a point G, that does not lie on the segment FC. By Axiom II, 4 the line EG must then intersect the segment AC at a point D.

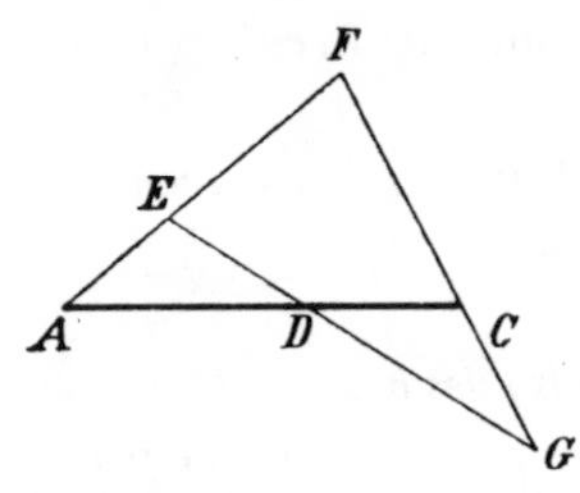

THEOREM 4. Of any three points A, B, C on a line there always is one that lies between the other two.

PROOF.[1] Let A not lie between B and C and let also C not lie between A and B. Join a point D that does not lie on the line AC with B and choose by Axiom II, 2 a point G on the connecting line such that D lies between B and G. By an application of Axiom II, 4 to the triangle BCG and to the line AD it follows that the lines AD and CG intersect at a point E that lies between C and G. In the same way, it follows that the lines CD and AG meet at a point F that lies between A and G.

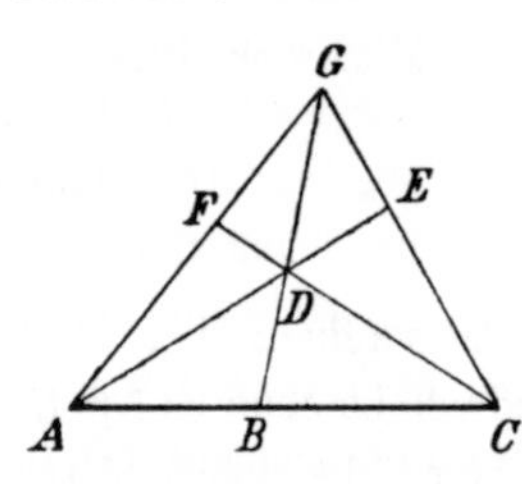

If Axiom II, 4 is applied now to the triangle AEG and to the line CF it becomes evident that D lies between A and E, and by an application of the same axiom to the triangle AEC and to the line BG one realizes that B lies between A and C.

THEOREM 5. Given any four points on a line, it is always possible to label them A, B, C, D in such a way that the point labeled B lies between A and C and also between A and D, and furthermore, that the point labeled C lies between A and D and also between B and D.[2]

[1] This proof is due to A. Wald.

[2] This theorem, which had been given in the First Edition as an axiom, was recognized by E. H. Moore, *Trans. Am. Math. Soc.*, 1902, to be a consequence of the plane axioms of incidence and order formulated above. Compare also the works subsequent to this by Veblen, *Trans. Am. Math. Soc.*, 1904, and Schweitzer, *American Journal*, 1909. A thorough investigation of independent sets of line axioms of order that postulate ordering on straight lines is found in E. v. Huntington, "A New Set of Postulates for Betweenness with Proof of Complete Independence," *Trans. Am. Math. Soc.*, 1924. Compare also *Trans. Am. Math. Soc.*, 1917.

PROOF. Let A, B, C, D be four points on a line g. The following will now be shown:

1. If B lies on the segment AC and C lies on the segment BD then the points B and C also lie on the segment AD. By Axioms I, 3 and II, 2 choose a point E that does not lie on g, on a point F such that E lies between C and F. By repeated applications of Axioms II, 3 and II, 4 it follows that the segments AE and BF meet at a point G, and moreover, that the line CF meets the segment GD at a point H. Since H thus lies on the segment GD and since, however, by Axiom II, 3, E does not lie on the segment AG, the line EH, by Axiom II, 4, meets the segment AD, i.e., C lies on the segment AD. In exactly the same way one shows analogously that B also lies on this segment.

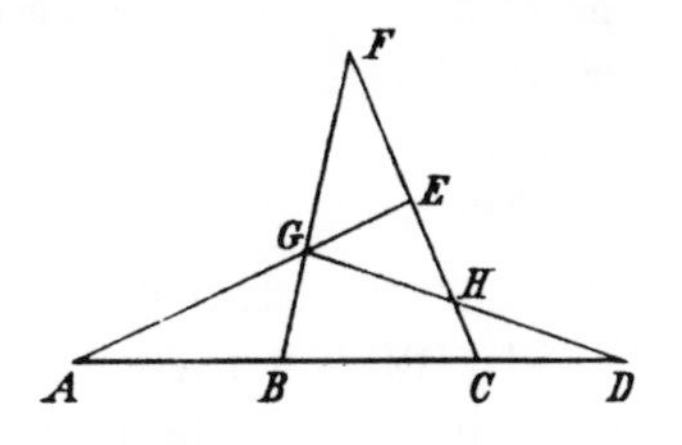

2. If B lies on the segment AC and C lies on the segment AD then C also lies on the segment BD and B also lies on the segment AD. Choose one point G that does not lie on g, and another point F such that G lies on the segment BF. By Axioms I, 2 and II, 3 the line CF meets neither the segment AB nor the segment BG and hence, by Axiom II, 4 again, does not meet the segment AG. But since C lies on the segment AD, the straight line CF meets then the segment GD at a point H. Now by Axiom II, 3 and II, 4 again the line FH meets the segment BD. Hence C lies on the segment BD. The rest of Assertion 2 thus follows from 1.

Now let any four points on a line be given. Take three of the points and label Q the one which by Theorem 4 and Axiom II, 3 lies between the other two and label the other two P and R. Finally label S the last of the four points. By Axiom II, 3 and Theorem 4 again it follows then that the following five distinct possibilities for the position of S exist:

R lies between P and S,

or P lies between R and S,

or S lies between P and R simultaneously when Q lies between P and S,

or S lies between P and Q,

or P lies between Q and S.

The first four possibilities satisfy the hypotheses of 2 and the last one satisfies those of 1. Theorem 5 is thus proved.

THEOREM 6 (generalization of Theorem 5). Given any finite number of points on a line it is always possible to label them $A, B, C,$

$D, E, \ldots, K$ in such a way that the point labeled B lies between A and $C, D, E, \ldots, K$, the point labeled C lies between A, B

and $D, E, \ldots, K$, D lies between A, B, C and $E, \ldots, K$, etc. Besides this order of labeling there is only the reverse one that has the same property.

THEOREM 7. Between any two points on a line there exists an infinite number of points.

THEOREM 8. Every line a that lies in a plane α separates the points which are not on the plane α into two regions with the following property: Every point A of one region determines with every point B of the other region a segment AB on which there lies a point of the line a. However any two points A and A' of one and the same region determine a segment AA' that contains no point of a.

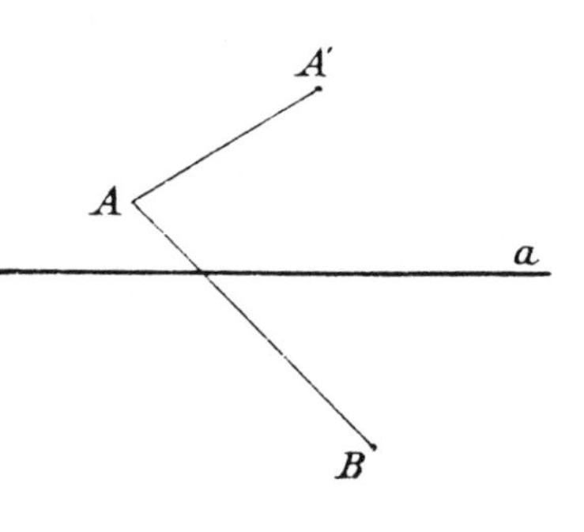

DEFINITION. The points A, A' are said to lie *in the plane α on one and the same side of the line a* and the points A, B are said to lie *in the plane α on different sides of the line a.*

DEFINITION. Let A, A', O, B be four points of the line a such that O lies between A and B but not between A and A'. The points A, A' are then said to lie *on the line a on one and the same side of the point O* and the points A, B are said to lie *on the line a on different sides of the point O.* The totality of points of the line a that lie on one and the same side of O is called a *ray* emanating from O. Thus every point of a line partitions it into two rays.

DEFINITION. A set of segments $AB, BC, CD, \ldots, KL$ is called a *polygonal segment* that connects the points A and L. Such a polygonal segment will also be briefly denoted by $ABCD \ldots KL$. The points inside the segments $AB, BC, CD, \ldots, KL$ as well as the points $A, B, C, D, \ldots, K, L$ are collectively called the *points of the polygonal segment*. If the points $A, B, C, D, \ldots, K, L$ all lie in a plane and the point A coincides with the point L then the polygonal segment is called a *polygon* and is denoted as the polygon $ABCD \ldots K$. The segments $AB, BC, CD, \ldots, KA$ are also called the *sides of the polygon*. The points $A, B, C, D, \ldots, K$ are called the *vertices of the polygon*. Polygons of 3, 4, . . . , *n vertices* are called *triangles, quadrilaterals, . . . , n-gons*.

DEFINITION. If the vertices of a polygon are all distinct, none of them falls on a side and no two of its nonadjacent sides have a point in common, the polygon is called *simple*.

With the aid of Theorem 8 the following theorems can now be obtained:* (See the bibliography at the end of Supplement I, 1.)

THEOREM 9. Every single polygon lying in a plane α separates the points of the plane α that are not on the polygonal segment of the polygon into two regions, the *interior* and the *exterior*, with the following property: If A is a point of the interior **(an inner point)** and B is a point of the exterior **(an exterior point)** then every polygonal segment that lies in α and joins A with B has at least one point in common with the polygon. On the other hand if A, A' are two points of the interior and B, B' are two points of the exterior then there exist polygonal segments in α which join A with A' and others which join B with B', none of which have any point in common with the polygon. By suitable labeling of the two regions there exist lines in α that always lie entirely in the exterior of the polygon. However, there are no lines that lie entirely in the interior of the polygon.

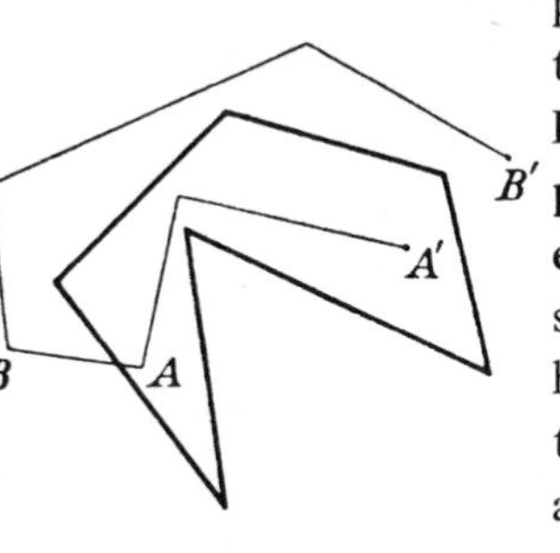

THEOREM 10. Every plane α separates the other points of space into two regions with the following property: Every point A of one region determines with every point B of the other region a segment AB on which there lies a point of α; whereas, two points A and A' of one and the same region always determine a segment AA' that contains no point of α.

DEFINITION. In the notation of Theorem 10 it is said that the points A, A' lie in space *on one and the same side of the plane* α and the points A, B lie in space *on different sides of the plane* α.

Theorem 10 expresses the most important facts about the ordering of the elements of **space**. These facts are thus merely consequences of

*This sentence is one of the corrections alluded to in the preface to the Ninth Edition of this book. Hilbert's original statement, which was retained in previous editions, was "With the aid of Theorem 8 one obtains without much difficulty the following theorem." (Translator's note)

the axioms considered so far and thus no new **space** axiom is required in Group II.

§ 5. Axiom Group III: Axioms of Congruence

The axioms of this group define the concept of congruence and with it also that of displacement.

DEFINITION. Segments stand in a certain relation to each other and for its description the words "*congruent*" or "*equal*" will be used.

III, 1. *If A, B are two points on a line a, and A' is a point on the same or on another line a' then it is always possible to find a point B' on a given side of the line a' through A' such that the segment AB is congruent or equal to the segment $A'B'$. In symbols*

$$AB \equiv A'B'.$$

This axiom requires the **possibility of** *constructing* **segments. Its uniqueness** will be proved later.

A segment was simply defined as a set of two points A, B and was denoted by AB or BA. In the definition the order of the two points was not specified. Therefore, the formulas

$$AB \equiv A'B', \qquad AB \equiv B'A',$$
$$BA \equiv A'B', \qquad BA \equiv B'A'$$

have equal meanings.

III, 2. *If a segment $A'B'$ and a segment $A''B''$, are congruent to the same segment AB, then the segment $A'B'$ is also congruent to the segment $A''B''$, or briefly, if two segments are congruent to a third one they are congruent to each other.*

Since congruence or equality is introduced in geometry only through these axioms, it is by no means obvious that **every segment is congruent to itself.** However, this fact follows from the first two axioms on congruence if the segment AB is constructed on a ray so that it is congruent, say, to $A'B'$ and Axiom III, 2 is applied to the congruences $AB \equiv A'B'$, $AB \equiv A'B'$.

On the basis of this the *symmetry* and the *transitivity* of segment congruence can be established by an application of Axiom III, 2; i.e., the validity of the following theorems:

If $$AB \equiv A'B',$$
then $$A'B' \equiv AB;$$
if $$AB \equiv A'B'$$

and $\qquad A'B' \equiv A''B''$,

then $\qquad AB \equiv A''B''$.

Due to the symmetry of segment congruence one may use the expression "Two segments are *congruent to each other.*"

III, 3. *On the line a let AB and BC be two segments which except for B have no point in common. Furthermore, on the same or on another line a′ let*

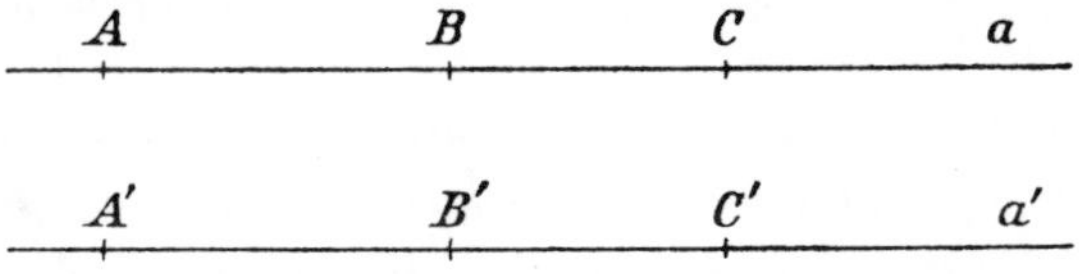

A B C a

A′ B′ C′ a′

A′B′ and B′C′ be two segments which except for B′ also have no point in common. In that case, if

$$AB \equiv A'B' \text{ and } BC \equiv B'C'$$

then $$AC \equiv A'C'.$$

This axiom expresses the requirement of **additivity** of segments.

The construction of angles is dealt with precisely as the construction of segments. Besides the **possibility** of constructing angles it is by all means also necessary to require **uniqueness** axiomatically. However, transitivity and additivity can be proved.

DEFINITION. Let α be a plane and h, k any two distinct rays emanating from O in α and lying on **distinct lines.** The pair of rays h, k is called an *angle* and is denoted by $\measuredangle(h, k)$ or by $\measuredangle(k, h)$.

The rays h, k are called the *sides* of the angle and the point O is called the *vertex* of the angle.

Degenerate and obtuse angles are excluded by this definition.

Let the ray h lie on the line h and the ray k on the line $\bar{k}$. The rays h and k together with the point O partition the points of the plane into two regions. All points that lie on the same side of k as those on h, and also those that lie on the same side of $\bar{h}$ as those on k, are said to lie in the **interior** of the angle $\measuredangle(h, k)$. All other points are said to lie in the **exterior** of, or outside, this angle.

It is easy to see by Axioms I and II that both regions contain points and that a segment that connects two points **inside** the angle lies entirely in the interior. The following facts are just as easy to prove: If a point H lies on h and a point K lies on k then the segment HK lies entirely in the interior. A ray emanating from O lies either entirely inside or entirely outside the angle. A ray that lies in the interior meets the segment HK. If A is a point of one region and B is a point of the other region, then every polygonal segment that connects A and B either passes through O or has at least one point in common with h or with k. However, if A, A' are points of the same region then there always exists a polygonal segment that connects A with A' and passes neither through O nor through any point of the rays h, k.

DEFINITION. Angles stand in a certain relation to each other, and for the description of which the word "*congruent*" or "*equal*" will be used.

III, 4. *Let $\sphericalangle(h, k)$ be an angle in a plane α and a' a line in a plane α' and let a definite side of a' in α' be given. Let h' be a ray on the line a' that emanates from the point O'. Then there exists in the plane α' one and only one ray k' such that the angle $\sphericalangle(h, k)$ is congruent or equal to the angle $\sphericalangle(h', k')$ and at the same time all interior points of the angle $\sphericalangle(h', k')$ lie on the given side of a'.*
Symbolically

$$\sphericalangle(h, k) \equiv \sphericalangle(h', k').$$

Every angle is congruent to itself, i.e.,

$$\sphericalangle(h, k) \equiv \sphericalangle(h, k).$$

is always true.

One also says briefly that every angle in a given plane can be *constructed* on a given side of a given ray in a uniquely determined way.

In the definition of an angle just as little consideration will be given to its orientation as has been given to the sense of a segment. Consequently the designations $\sphericalangle(h, k)$, $\sphericalangle(k, h)$ will have the same meaning.

DEFINITION. An angle with a vertex B on one of whose sides lies a point A and on whose other side lies a point C will also be denoted by $\sphericalangle ABC$ or briefly by $\sphericalangle B$. Angles will also be denoted by small Greek letters.

III, 5. *If for two triangles*[1] *ABC and $A'B'C'$ the congruences*

$$AB \equiv A'B', \quad AC \equiv A'C', \quad \sphericalangle BAC \equiv \sphericalangle B'A'C'$$

hold, then the congruence

$$\sphericalangle ABC \equiv \sphericalangle A'B'C'$$

is also satisfied.

[1] Here, and in what follows, the vertices of a triangle shall always be supposed not to lie on the same line.

The concept of a triangle is defined on p. 9. Under the hypotheses of the axiom it follows, by a change of notation, that **both congruences**

$$\sphericalangle ABC \equiv \sphericalangle A'B'C' \text{ and } \sphericalangle ACB \equiv \sphericalangle A'C'B'$$

are satisfied.

Axioms III, 1-3 contain statements about the congruence of segments. They may therefore be called the *line* axiom of group III. Axiom III, 4 contains statements about the congruence of angles. Axiom III, 5 relates the concepts of congruence of segments to that of angles. Axioms III, 4 and III, 5 contain statements about the elements of plane geometry and may therefore be called the *plane* axioms of group III.

The **uniqueness of segment construction** follows from the uniqueness of angle construction with the aid of Axiom III, 5.

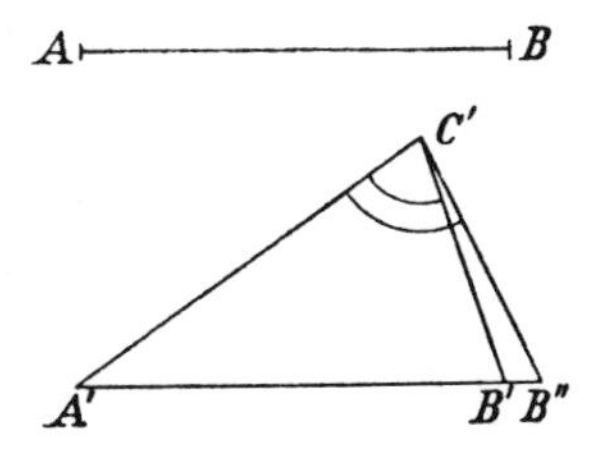

Suppose that the segment AB is constructed in two ways on a ray emanating from A' to B' and to B''. Choosing a point C' not on the line $A'B'$ the congruences

$$A'B' \equiv A'B'', \quad A'C' \equiv A'C', \quad \sphericalangle B'A'C' \equiv \sphericalangle B''A'C',$$

are obtained and so by Axiom III, 5

$$\sphericalangle A'C'B' \equiv \sphericalangle A'C'B'',$$

in contradiction to the uniqueness of angle construction required by Axiom III, 4.

§ 6. Consequences of the Axioms of Congruence

DEFINITION. Two angles that have a vertex and one side in common and whose separate sides form a line are called *supplementary angles*. Two angles with a common vertex whose sides form two lines are called *vertical angles*. An angle that is congruent to one of its supplementary angles is called a *right angle*.

The following theorems will now be proved:

THEOREM 11. In a triangle the angles opposite two congruent sides are congruent, or briefly, the base angles of an isosceles triangle are equal.

This theorem follows from Axiom III, 5 and the last part of Axiom III, 4.

DEFINITION. A triangle ABC is said to be congruent to a triangle $A'B'C'$ if all congruences

$$AB \equiv A'B', \quad AC \equiv A'C', \quad BC \equiv B'C'$$

$$\measuredangle A \equiv \measuredangle A', \quad \measuredangle B \equiv \measuredangle B', \quad \measuredangle C \equiv \measuredangle C'$$

are satisfied.

THEOREM 12. **(first congruence theorem for triangles)**. A triangle ABC is congruent to a triangle $A'B'C'$ whenever the congruences

$$AB \equiv A'B', \quad AC \equiv A'C', \quad \measuredangle A \equiv \measuredangle A'$$

hold.

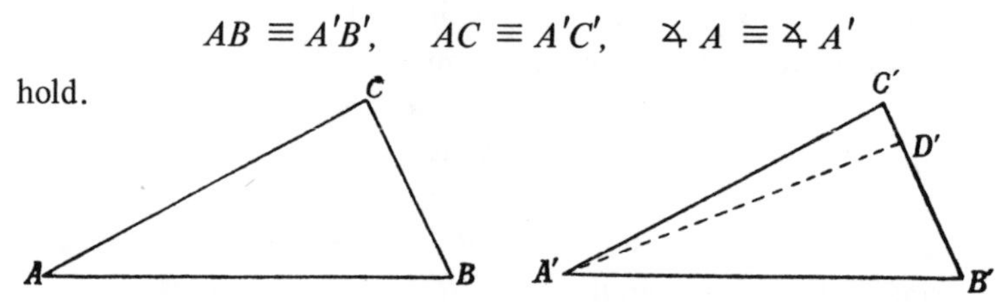

PROOF. By Axiom III, 5 the congruences

$$\measuredangle B \equiv \measuredangle B' \text{ and } \measuredangle C \equiv \measuredangle C'$$

are satisfied and thus it is only necessary to prove the validity of the congruence $BC \equiv B'C'$. If it is assumed to the contrary that BC is not congruent to $B'C'$ and a point D' is determined on $B'C'$ so that $BC \equiv B'D'$ then Axiom III, 5, applied to both triangles ABC and $A'B'D'$, will indicate that $\measuredangle BAC \equiv \measuredangle B'A'D'$. Then $\measuredangle BAC$ would be congruent to $\measuredangle B'A'D'$ as well as to $\measuredangle B'A'C'$. This is impossible, as by Axiom III, 4 every angle can be constructed on a given side of a given ray in a plane in only **one** way. It has thus been proved that the triangle ABC is congruent to the triangle $A'B'C'$.

The following is just as easy to prove:

THEOREM 13 **(second congruent theorem for triangles)**. A triangle ABC is congruent to another triangle $A'B'C'$ whenever the congruences

$$AB \equiv A'B', \quad \measuredangle A \equiv \measuredangle A', \quad \measuredangle B \equiv \measuredangle B'$$

hold.

THEOREM 14. If an angle $\measuredangle ABC$ is congruent to another angle

$\measuredangle\, A'B'C'$ then its supplementary angle $\measuredangle\, CBD$ is congruent to the supplementary angle $\measuredangle\, C'B'D'$ of the other angle.

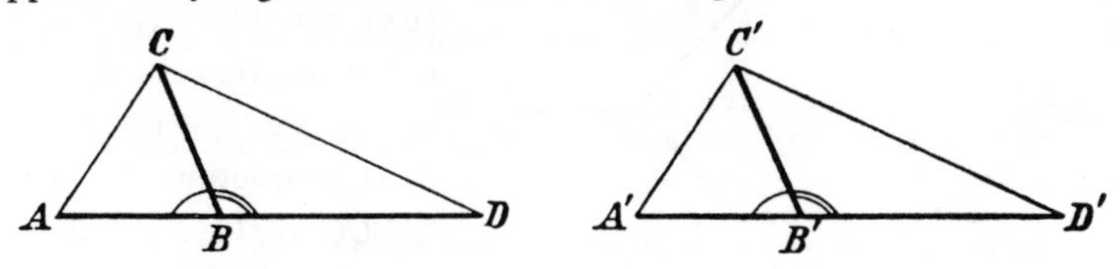

PROOF. Choose the points A', C', D' on the sides that pass through B' in such a way that

$$AB \equiv A'B', \quad CB \equiv C'B', \quad DB \equiv D'B'.$$

It follows then from Theorem 12 that the triangle ABC is congruent to the triangle $A'B'C'$, i.e., the congruences

$$AC \equiv A'C' \text{ and } \measuredangle\, BAC \equiv \measuredangle\, B'A'C'$$

hold.

Since moreover by Axiom III, 3 the segment AD is congruent to the segment $A'D'$, it follows again by Theorem 12 that the triangle CAD is congruent to the triangle $C'A'D'$, i.e., the congruences

$$CD \equiv C'D' \text{ and } \measuredangle\, ADC \equiv \measuredangle\, A'D'C'$$

hold and hence, by considering the triangles BCD and $B'C'D'$, it follows by Axiom III, 5 that

$$\measuredangle\, CBD \equiv \measuredangle\, C'B'D'.$$

An immediate corollary of Theorem 14 is the **congruence theorem for vertical angles.**

The **existence of right angles** also follows from this theorem (see p. 15).

If angles constructed on both sides of a ray OA emanating from O and if the two noncommon sides of the angle are made equal, $OB \equiv OC$, then the segment BC intersects the line OA at a point D. If D

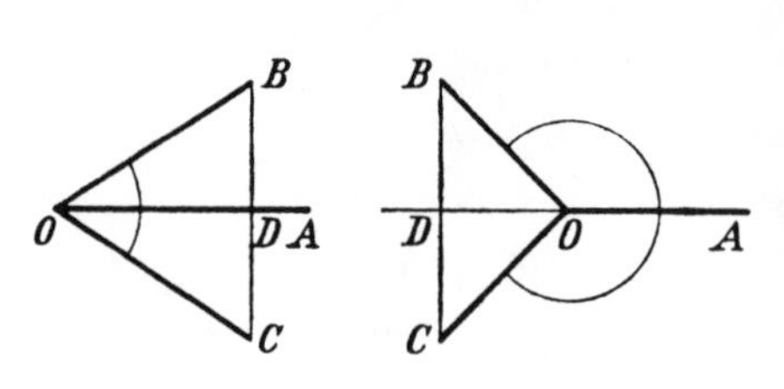

coincides with O then $\measuredangle COA$ and $\measuredangle BOA$ are equal supplementary angles and hence are right angles. If D lies on the ray OA then by the construction $\measuredangle DOB \equiv \measuredangle DOC$. If D lies on the other ray then the congruence follows from Theorem 14. By Axiom III, 2 every segment is congruent to itself: $OD \equiv OD$. Hence, by Axiom III, 5, it follows that $\measuredangle ODB \equiv \measuredangle ODC$.

THEOREM 15. Let h, k, l and h', k', l' be rays emanating from O and O' in the planes α and α', respectively. Let h, k and h', k' lie simultaneously on the same or on different sides of l and l', respectively. If the congruences

$$\measuredangle (h, l) \equiv \measuredangle (h', l') \quad \text{and} \quad \measuredangle (k, l) \equiv \measuredangle (k', l')$$

are satisfied then so is the congruence

$$\measuredangle (h, k) \equiv \measuredangle (h', k').$$

The **proof** will be given for the case when h and k lie on the same side of l and hence by the hypothesis when h' and k' lie on the same side of l'. The second case will be reduced to the first case by an application of Theorem 14. From the definition on p. 11 it follows that either h lies in the angle $\measuredangle (k, l)$ or that k lies in the angle $\measuredangle (h, l)$. Now label so that h lies in the angle $\measuredangle (k, l)$. Choose the points K, K', L, L' on the sides k, k', l, l' so that $OK \equiv O'K'$ and $OL \equiv O'L'$. By a theorem stated on p. 11 h intersects the segment KL at a point H.

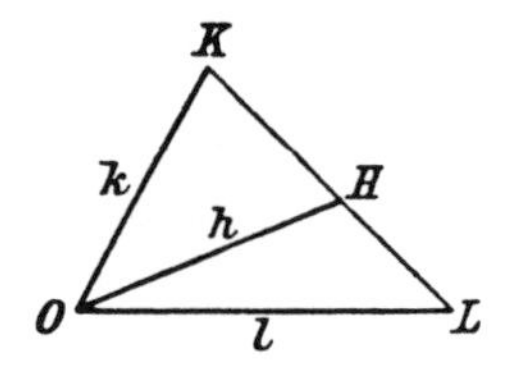

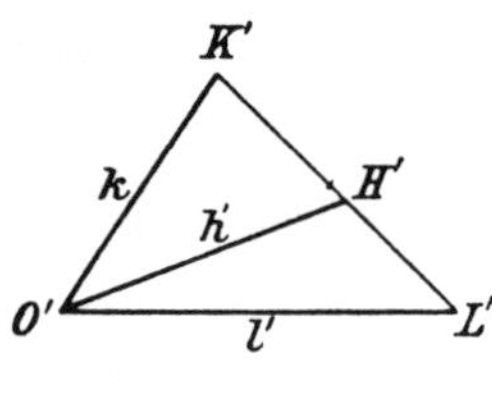

Determine H' on h' so that $OH \equiv O'H'$. By Theorem 12 the congruences $\measuredangle OLH \equiv \measuredangle O'L'H'$, $\measuredangle OLK \equiv \measuredangle O'L'K'$,

$$LH \equiv L'H', \qquad LK \equiv L'K'$$

and also

$$\measuredangle OKL \equiv \measuredangle O'K'L'$$

are obtained in the triangles OLH and $O'L'H'$ or OLK and $O'L'K'$.

Since by Axiom III, 4 every angle can be constructed on a given side of a given ray in a plane in only one way and since by hypothesis

H' and K' lie on the same side of l' the first two mentioned angle congruences show that H' lies on $L'K'$. Hence the two segment congruences show easily by Axiom III, 3 and the uniqueness of segment construction that $HK \equiv H'K'$. The assertion is now deduced by Axiom III, 5 from the congruence $OK \equiv O'K'$, $HK \equiv H'K'$ and $\sphericalangle OKL \equiv \sphericalangle O'K'L'$.

The following result is obtained in a similar way:

THEOREM 16. Let the angle $\sphericalangle (h, k)$ in the plane α be congruent to the angle $\sphericalangle (h', k')$ in the plane α', and let l be a ray in the plane α that emanates from the vertex of the angle $\sphericalangle (h, k)$ and which lies in the interior of this angle. Then there always exists one and only one ray l' in the plane α' that emanates from the vertex of the $\sphericalangle (h', k')$ and which lies in the interior of this angle in such a way that

$$\sphericalangle (h, l) \equiv \sphericalangle (h', l') \text{ and } \sphericalangle (k, l) \equiv \sphericalangle (k', l').$$

In order to obtain the third congruence theorem and the symmetry property of angle congruence the following theorem is now deduced from Theorem 15:

THEOREM 17. If two points Z_1 and Z_2 are placed on different sides of a line XY and if the congruences $XZ_1 \equiv XZ_2$ and $YZ_1 \equiv YZ_2$ hold, then the angle $\sphericalangle XYZ_1$ is congruent to the angle $\sphericalangle XYZ_2$.

PROOF. By Theorem 11 $\sphericalangle XZ_1Z_2 \equiv \sphericalangle XZ_2Z_1$ and $\sphericalangle YZ_1Z_2 \equiv \sphericalangle YZ_2Z_1$. Hence the congruence $\sphericalangle XZ_1Y \equiv \sphericalangle XZ_2Y$ follows from Theorem 15. The special cases when X or Y lies on Z_1Z_2 can be disposed of in an even simpler manner. From the last congruence and the assumed congruences $XZ_1 \equiv XZ_2$ and $YZ_1 \equiv YZ_2$ the assertion $\sphericalangle XYZ_1 \equiv \sphericalangle XYZ_2$ follows by Axiom III, 5.

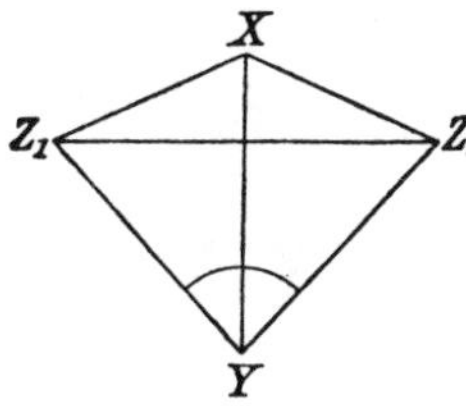

THEOREM 18 (**third congruence theorem for triangles**). If in two triangles ABC and $A'B'C'$ each pair of corresponding sides is congruent then so are the triangles.

PROOF. By virtue of the symmetry of segment congruence proved on p. 11 it is sufficient to prove that the triangle ABC is congruent to the triangle $A'B'C'$. Construct the angle $\sphericalangle BAC$ at A' on both sides of

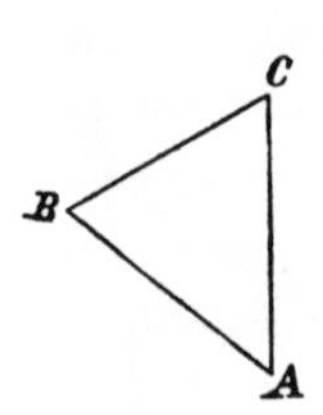

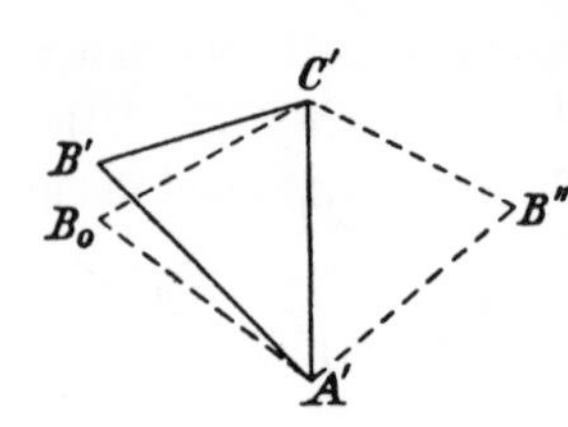

the ray $A'C'$. Choose the point B_0 on the side that lies on the same side of $A'C'$ as B' so that $A'B_0 \equiv AB$. On the other side let B'' be chosen in such a way that $A'B'' \equiv AB$. By Theorem 12 $BC \equiv B_0C'$ and $BC \equiv B''C'$. These congruences together with those in the hypothesis yield by Axiom III, 2 the congruences

$$A'B'' \equiv A'B_0, \quad B''C' \equiv B_0C'$$

and correspondingly

$$A'B'' \equiv A'B', \quad B''C' \equiv B'C'.$$

Both the triangles $A'B''C'$ and $A'B_0C'$ as well as the triangles $A'B''C'$ and $A'B'C'$ satisfy the hypotheses of Theorem 17, i.e., the angle $\measuredangle B''A'C'$ is congruent to the angle $\measuredangle B_0A'C'$ as well as the angle $\measuredangle B'A'C'$. But since by Axiom III, 4 every angle can be constructed on a given side of a given ray in a plane in only **one** way, the ray $A'B_0$ coincides with the ray $A'B'$, i.e., the angle that is congruent to $\measuredangle BAC$, constructed on the given side of $A'C'$, is the angle $\measuredangle B'A'C'$. The assertion follows then by Theorem 12 from the congruence $\measuredangle BAC \equiv \measuredangle B'A'C'$ and the assumed segment congruences.

THEOREM 19. If two angles $\measuredangle (h', k')$ and $\measuredangle (h'', k'')$ are congruent to a third angle $\measuredangle (h, k)$ then the angle $\measuredangle (h', k')$ is also congruent to angle $\measuredangle (h'', k'')$.[1]

This theorem which corresponds to Axiom III, 2 can also be formulated in this way. If two angles are congruent to a third one then they are congruent to each other.

PROOF. Let the vertices of the three given angles be O', O'' and O. On one side of each angle choose the points A', A'' and A so that $O'A' \equiv OA$ and $O''A'' \equiv OA$. Similarly, on the third sides choose the

[1] The proof given here for Theorem 19, which in the First Edition was taken as an axiom, is due to A. Rosenthal. Cf. *Math. Ann.*, Vol. 71.

The modified form of Axioms I, 3 and I, 4 is also due to A. Rosenthal. Cf. *Math. Ann.*, Vol. 69.

points B', B'' and B so that $O'B' \equiv OB$ and $O''B'' \equiv OB$. These congruences together with the assumption $\measuredangle(h', k') \equiv \measuredangle(h, k)$ and $\measuredangle(h'', k'') \equiv \measuredangle(h, k)$ yield by Theorem 12 the congruences

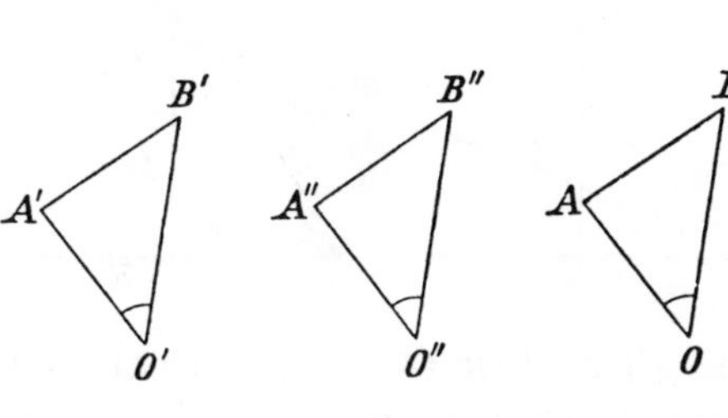

$$A'B' \equiv AB \quad \text{and} \quad A''B'' \equiv AB.$$

By Axiom III, 2 the triangles $A'B'O'$ and $A''B''O''$ coincide in their three sides and hence by Theorem 18

$$\measuredangle(h', k') \equiv \measuredangle(h'', k'').$$

The symmetry property of angle congruence follows from Theorem 19 just as it does for segments from Axiom III, 2; i.e., if $\measuredangle\alpha \equiv \measuredangle\beta$ then $\measuredangle\alpha$ and $\measuredangle\beta$ are **congruent to each other**. In particular Theorems 12-14 can be expressed now in symmetric form.

The **quantitative comparison of angles** can now be established.

THEOREM 20. Let any two angles $\measuredangle(h, k)$ and $\measuredangle(h', l')$ be given. If the construction of $\measuredangle(h, k)$ on h' on the side of l' yields an **interior** ray k' then the construction of $\measuredangle(h', l')$ on h on the side of k yields an **exterior** ray l, and conversely.

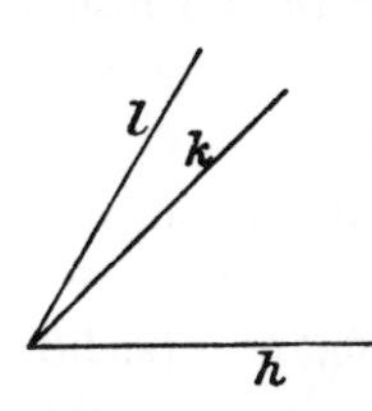

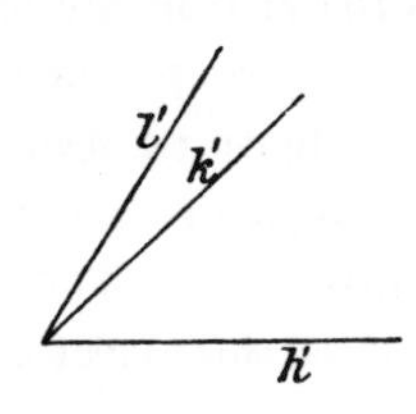

PROOF. By hypothesis l lies in the interior of $\measuredangle(h, k)$. Since $\measuredangle(h, k) \equiv \measuredangle(h', k')$ by Theorem 16 there exists for the interior ray l a ray l'' in the **interior** of $\measuredangle(h', k')$ for which the congruence $\measuredangle(h, l) \equiv \measuredangle(h', l'')$ holds. By hypothesis and by virtue of the symmetry of angle congruence $\measuredangle(h, l) \equiv \measuredangle(h', l')$ where l' and l'' are necessarily

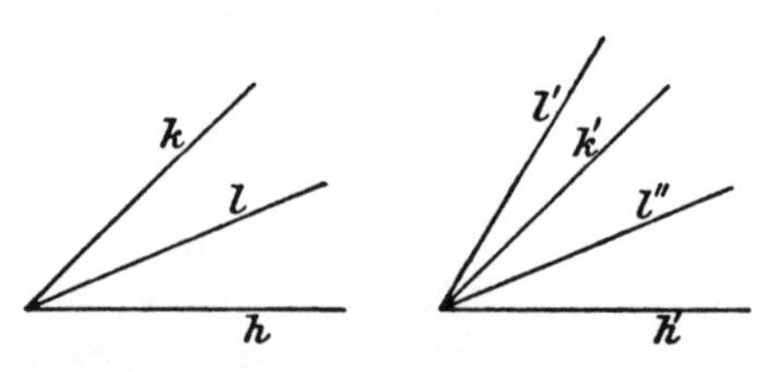

distinct, contrary to the uniqueness of angle construction III, 4. The converse is proved similarly.

If the construction of $\measuredangle (h, k)$ described in Theorem 20 yields an interior ray k' in $\measuredangle (h', l')$ it is said that $\measuredangle (h, k)$ is *smaller than* $\measuredangle (h', l')$; symbolically $\measuredangle (h, k) < \measuredangle (h', l')$. If it yields an exterior ray it is said that $\measuredangle (h, k)$ *is greater than* $\measuredangle (h', l')$; symbolically $\measuredangle (h, k) > \measuredangle (h', l')$.

It should be realized that for two angles α and β **one and only one of the three cases**

$$\alpha < \beta \text{ and } \beta > \alpha, \quad \alpha \equiv \beta, \quad \alpha > \beta \text{ and } \beta < \alpha$$

can exist. The quantitative comparison of angles is *transitive*, i.e., from each of the three assumptions

$$1.\ \alpha > \beta,\ \beta > \gamma; \quad 2.\ \alpha > \beta,\ \beta \equiv \gamma; \quad 3.\ \alpha \equiv \beta,\ \beta > \gamma$$

follows

$$\alpha > \gamma.$$

The **quantitative comparison of segments** with corresponding properties follows immediately from Axioms II and III, 1-3 as well as from the uniqueness of segment construction (see p. 13).

On the basis of the quantitative comparison of angles it is possible to obtain a proof for the following simple theorem which, subjectively speaking, *Euclid* listed unjustifiedly among the axioms:

THEOREM 21. *All right angles are congruent to each other.*

PROOF.[1] By definition a right angle is one that is congruent to its complementary angle. Let the angles α or $\measuredangle (h, l)$ and β or $\measuredangle (k, l)$ be supplementary angles and let α' and β' also be such angles. Let $\alpha \equiv \beta$ and $\alpha' \equiv \beta'$. Suppose that α' is not congruent to α, contrary to the hypothesis of Theorem

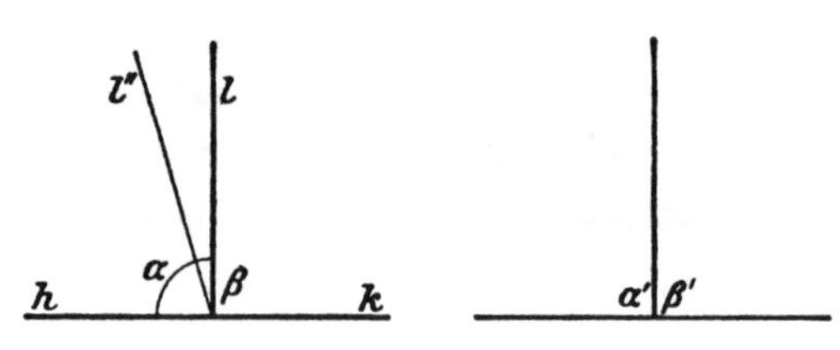

[1] The idea of this proof can be found as early as the Euclid commentator **Proclus**, who indeed, instead of Theorem 14 used the hypothesis that the construction of one right angle always yields another right angle, i.e., yields an

21. Then the construction of the angle α' on h on the side on which l lies yields a ray l'' that is distinct from l. l'' lies then either in the interior of α or in the interior of β. If l'' lies in the interior of α then

$$\sphericalangle(h, l'') < \alpha, \ \alpha \equiv \beta, \ \ \beta < \sphericalangle(k, l'').$$

By the transitivity of the quantitative comparison of angles it follows from this that $\sphericalangle(h, l'') < \sphericalangle(k, l'')$. On the other hand, by the hypothesis and by Theorem 14

$$\sphericalangle(h, l'') \equiv \alpha', \ \alpha' \equiv \beta', \ \ \beta' \equiv \sphericalangle(k, l''),$$

and hence follows

$$\sphericalangle(h, l'') \equiv \sphericalangle(k, l''),$$

contrary to the relation $\sphericalangle(h, l'') < \sphericalangle(k, l'')$. If l'' lies in the interior of β a completely analogous contradiction is obtained and Theorem 21 is thus proved.

DEFINITION. An angle that is greater than its supplementary angle is called an *obtuse* angle. An angle that is smaller than its supplementary angle is called an *acute* angle.

A fundamental theorem that already played an important role for Euclid and from which follows a series of important results is the theorem of the exterior angle.

DEFINITION. The angles $\sphericalangle ABC$, $\sphericalangle BCA$ and $\sphericalangle CAB$ of the triangle ABC are called the *interior angles* of the triangle. Their supplementary angles are called the *exterior angles* of the triangle.

THEOREM 22 (**theorem of the exterior angle**). The exterior angle of a triangle is greater than any interior angle that is not adjacent to it.

PROOF. Let $\sphericalangle CAD$ be an exterior angle of the triangle ABC. D may be chosen so that $AD \equiv CB$.

It will be shown next that $\sphericalangle CAD \not\equiv \sphericalangle ACB$. If $\sphericalangle CAD \equiv \sphericalangle ACB$ held then so would $\sphericalangle ACD \equiv \sphericalangle CAB$ by virtue of the congruence $AC \equiv CA$ and by Axiom III, 5. It would follow from Theorems 14 and 19 that $\sphericalangle ACD$ would be congruent to the supplementary angle of $\sphericalangle ACB$. By Axiom III, 4 D would thus lie on the

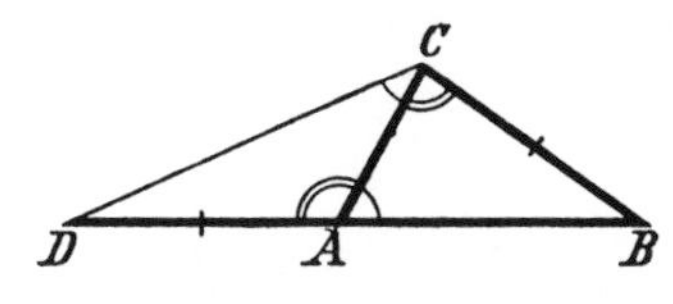

angle that is equal to its supplementary angle.

A French translation of the commentary of **Proclus** with an introduction and notes by P. Ver Eecke, "Proclus de Lycie"–Les Commentaires sur le premier livre des éléments d'Euclide, was published in *Collection de travaux de l'Acad. internat. d'histoire des sciences*, No. 1 (Brügge, 1948).

line CB, contrary to Axiom I, 2. It must then be that

$$\measuredangle CAD \not\equiv \measuredangle ACB.$$

It is also impossible that $\measuredangle\ CAD < \measuredangle\ ACB$ since then the construction of the exterior angle $\measuredangle CAD$ on CA at C on the side on which B lies would yield a side that lies in the interior of the angle $\measuredangle ACB$, and thus would meet the segment AB at point B'. The exterior angle $\measuredangle\ CAD$ would then be congruent to the angle $\measuredangle\ ACB'$ in the triangle $AB'C$. This however, as shown above, is impossible. It remains then only the possibility

$$\measuredangle\ CAD > \measuredangle\ ACB.$$

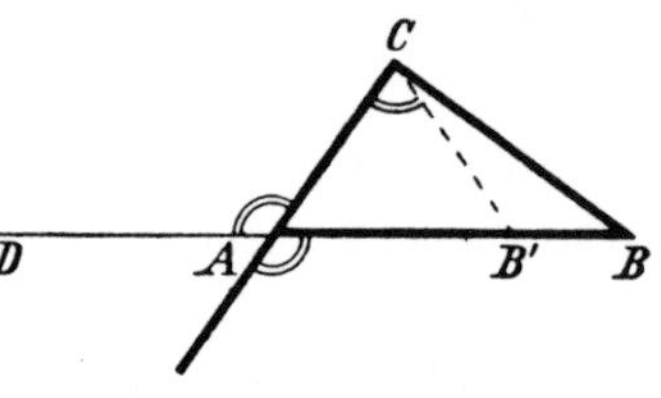

In exactly the same way one obtains the fact that the vertical angle of the angle $\measuredangle\ CAD$ is greater than the angle $\measuredangle\ ABC$, and from the congruence of vertical angles and the transitivity of the quantitative comparison of angle sizes it follows that

$$\measuredangle CAD > \measuredangle ABC.$$

The assertion is thus completely proved.

Important corollaries from this theorem are the following theorems:

THEOREM 23. In every triangle the greater angle lies opposite the greater side.

PROOF. In the given triangle construct the smaller of two sides with common end points on the greater one. The assertion follows then from Theorems 11 and 12 by virtue of the transitivity of the quantitative comparison of angle sizes.

THEOREM 24. A triangle with two equal angles is isosceles.

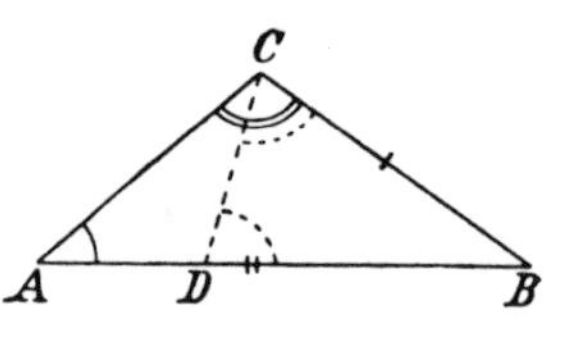

This converse of Theorem 11 is an immediate consequence of Theorem 23.

Moreover, from Theorem 22 follows in a simple way an extension to the second congruence theorem for triangles.

THEOREM 25. Two triangles ABC and $A'B'C'$ are congruent to each other if the congruences.

$$AB \equiv A'B' \quad \measuredangle A \equiv \measuredangle A' \text{ and } \measuredangle C \equiv \measuredangle C'$$

are satisfied.

THEOREM 26. Every segment can be bisected.

PROOF. On different sides of the given segment AB construct the same angle α at its end points and lay off equal segments on the third sides of the angles so that $AC \equiv BD$. Since C and D lie on different sides of AB the segment CD meets the line AB at a point E.

The assumption that E coincides with A or with B is an immediate contradiction of Theorem 22. Let it then be assumed that B lies between A and E. By Theorem 22 it would follow then that

$$\measuredangle ABD > \measuredangle BED > \measuredangle BAC,$$

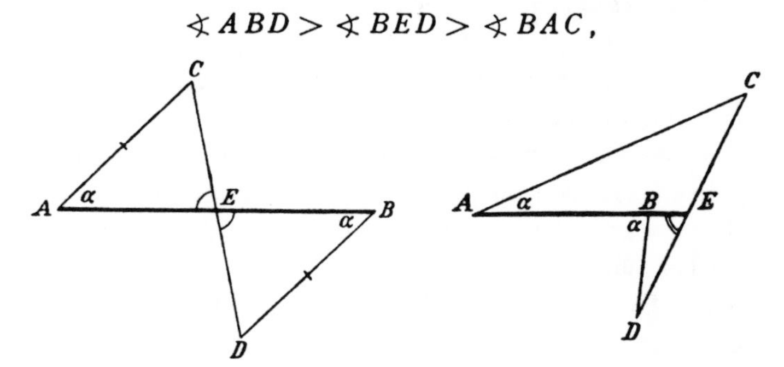

contrary to the construction. The assumption that A lies between B and E yields the same contradiction.

By Theorem 4 E lies then on the segment AB. Therefore $\measuredangle AEC$ and $\measuredangle BED$, as vertical angles, are congruent. Hence Theorem 25 is applicable to the triangles AEC and BED and yields

$$AE \equiv EB.$$

An immediate consequence of Theorems 11 and 26 is the fact that every angle can be bisected.

The concept of congruence can be extended to any figure.

DEFINITION. If $A, B, C, D, \ldots, K, L$ and $A', B', C', D', \ldots, K', L'$ are two sequences of points on a and a', respectively, such that the segments AB and $A'B'$, AC and $A'C'$, BC and $B'C'$, $\ldots$, KL and $K'L'$ are congruent to each other in pairs the two sequences of points are said to be congruent to each other. A and A', B and B', $\ldots$, L and L' are called the corresponding points of the *congruent sequences of points.*

THEOREM 27. If the first of the two sequences of congruent points $A, B, \ldots, K, L$ and $A', B', \ldots, K', L'$ is so ordered that B lies between A and $C, D, \ldots, K, L$; C lies between A, B, and $D, \ldots, K, L$, etc.; then the points $A', B', \ldots, K', L'$ are ordered in the same way, i.e., B' lies between A' and $C', D', \ldots, K', L'$; C' lies between A', B', and $D', \ldots, K', L'$, etc.

DEFINITION. A finite number of points is called a *figure*. If all points of a figure lie in a plane it is called a *plane figure*.

Two figures are said to be *congruent* if their points can be ordered in pairs so that the segments and the angles that become ordered in this way are all congruent to each other.

As becomes evident from Theorems 14 and 27 congruent figures have the following properties: If three points of a figure are collinear then the corresponding points in any congruent figure are also collinear. The ordering of points in corresponding planes with respect to corresponding lines is the same in congruent figures. The same holds for sequences of corresponding points on corresponding lines.

The most general congruence theorem for the plane and for space is expressed as follows:

THEOREM 28. If $(A, B, C, \ldots, L)$ and $(A', B', C', \ldots, L')$ are congruent plane figures and P denotes a point in the plane of the first figure then it is possible to find a point P' in the plane of the second figure so that $(A, B, C, \ldots, L, P)$ and $(A', B', C', \ldots, L', P')$ are again congruent figures. If the figure $(A, B, C, \ldots, L)$ contains at least three noncollinear points then the construction of P' is possible in only **one** way.

THEOREM 29. If $(A, B, C, \ldots, L)$ and $(A', B', C', \ldots, L')$ are congruent figures and P is any point, then it is always possible to find a point P' so that the figures $(A, B, C, \ldots, L, P)$ and $(A', B', C', \ldots, L', P')$ are congruent. If the figure $(A, B, C, \ldots, L)$ contains at least four noncoplanar points then the construction of P$'$ is possible in only **one** way.

By invoking axiom groups I and II, Theorem 29 expresses the important result that all **space** properties of congruence and thus the properties of displacement in **space** are consequences of the five **line** and **plane** axioms of congruence formulated above.

§ 7. Axiom Group IV: Axiom of Parallels

Let α be any plane, a any line in α and A a point in α not lying on a. If a line c is drawn in α so that it passes through A and intersects a, and a line b is drawn in α through A so that the line c intersects the lines a, b at the same angles then it follows easily from the exterior angle theorem, Theorem 22, that the lines a, b have no point in common, i.e., in a plane α it is always possible to draw a line through a point A not on a line a so that it does not intersect a.

DEFINITION. Two lines are said to be parallel if they lie in the same plane and do not intersect.

The axiom of parallels can be stated now as follows:

IV (**Euclid's Axiom**). *Let a be any line and A a point not on it. Then there is at most one line in the plane, determined by a and A, that passes through A and does not intersect a.*

From the foregoing and on the basis of the axiom of parallels it can be seen that there is exactly one parallel to a line through a point not on it.

The axiom of parallels IV is equivalent to the following requirement:

If two lines a, b in a plane do not meet a third line c in the same plane then they also do not meet each other.

In fact if a, b had a point A in common these two lines would pass through A in the same plane without meeting c. This situation would contradict the axiom of parallels IV. Conversely, the axiom of parallels also follows easily from this requirement.

The axiom of parallels is a *plane axiom*.

The introduction of the axiom of parallels **simplifies** the foundation of geometry and **facilitates** its development to a considerable degree.

Adjoining to the axioms of congruence the axiom of parallels the following familiar fact is obtained:

THEOREM 30. If two parallels are intersected by a third line then the corresponding and the alternate angles are congruent, and conversely, the congruence of the corresponding or the alternate angles implies that the lines are parallel.

THEOREM 31. The angles of a triangle add up to two right angles.[1]

[1] Concerning the question of how far this theorem can replace the converse of the axiom of parallels, compare the remarks in Section 12 at the end of Chapter II.

DEFINITION. If M is any point in a plane α then the collection of all points A in α for which the segments MA are congruent to each other is called a *circle*. M is called the *center of the circle*.

On the basis of this definition the familiar theorems about the circle follow easily with the aid of axiom groups III - IV—in particular, the possibility of constructing a circle through any three noncollinear points as well as the theorem about the congruence of inscribed angles over the same chord and the theorem of the angles in an inscribed quadrilateral.

§ 8. Axiom Group V: Axioms of Continuity

V, 1 (**Axiom of measure or Archimedes' Axiom**). *If AB and CD are any segments then there exists a number n such that n segments CD constructed contiguously from A, along the ray from A through B, will pass beyond the point B.*

V, 2 (**Axiom of line completeness**). *An extension of a set of points on a line with its order and congruence relations that would preserve the relations existing among the original elements as well as the fundamental properties of line order and congruence that follows from Axioms I-III, and from V, 1 is impossible.*

By the fundamental properties is meant the order properties formulated in Axioms II, 1-3 and in Theorem 5 as well as the congruence properties formulated in Axioms III, 1-3 along with the uniqueness of segment construction.[1] It is further meant that on extending the set of points the order and congruence relations carry over to the extended point region.

It should be noted that Axiom I, 3 is preserved at every extension *eo ipso* and that the validity of Theorem 3 at such extensions is a consequence of the persistence of Archimedes' Axiom V, 1.

The satisfaction of the axiom of completeness depends essentially on the fact that it contains Archimedes' Axiom among the axioms whose validity is required. In fact, it can be shown that to a set of points on a line that satisfies the previously enumerated axioms and theorems of order and congruence it is always possible to adjoin other points such that these axioms are also valid in the resulting extended

[1] A precise classification of the conditions to be required here for line ordering and congruence was carried out by F. Bachmann and was incorporated in the formulation of Axiom V, 2 in the Seventh Edition.

set; i.e., a completeness axiom that requires only the validity of these axioms but not that of Archimedes, or one that is equivalent to it, would entail a contradiction.

Both continuity axioms are *line* axioms.

The following fact is essentially obtained **from the line completeness axiom**:

THEOREM 32 (**theorem of completeness**).[1] The elements (i.e., the points, the lines and the planes) of geometry form a system which cannot be extended by points, lines and planes because of the persistence of the axioms of incidence, order, congruence and Archimedes, and so only because of the persistence of all axioms.[2]

PROOF. Let the elements that exist before the extension be designated as the **old** elements; those that arise from the extension be designated as the **new** elements. The assumption of new elements leads immediately to the assumption of a new point *N*.

By Axiom I, 8 there exist four old noncoplanar points *A, B, C, D*. The labels can be so chosen that *A, B, N* are not collinear. The two distinct planes *ABN* and *ACD* by Axiom I, 7 have besides *A*, another point *E* in common. *E* does not lie on the line *AB* for then *B* would lie in the plane *ACD*. If *E* is a new point then a new point *E* lies in the old plane *ACD*. On the other hand, if *E* is an old point then the new point *N* lies in an old plane, namely, in the plane *ABE*. In any case a new point lies in an old plane.

There exists an old triangle *FGH* in an old plane and on the segment *FG* and old point *I*. If a new point *L* is joined with *I* then by Axiom II, 4 the lines *IL* and *FH* or the lines *IL* and *GH* meet at a point *K*. If *K* is new then a new point *K* lies on an old line *FH* or *GH*. If on the other hand *K* is old then a new point *L* lies on an old line *IK*. All these assumptions are thus contrary to the axiom of line completeness. The assumption of a new point in an old plane must therefore be dropped and thereby the assumption of new elements.

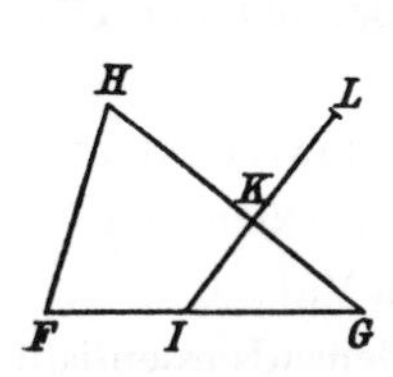

[1] The observation that the line completeness axiom is sufficient is due to P. Bernays.

[2] This assertion was stated in the previous editions as an axiom of completeness.

The completeness theorem can even be sharpened. The persistence of some of the mentioned axioms need not be unconditionally required for it. However, essential for its validity is that Axiom I, 7 be contained among the axioms whose persistence it requires. In fact it can be shown that to a set of elements which satisfies Axioms I-V it is always possible to adjoin points, lines and planes so that these same axioms with the exception of Axiom I, 7 hold in the set that arises from the adjunction, i.e., a completeness theorem in which Axiom I, 7 or one that is equivalent to it, is not contained would entail a contradiction.

The **completeness axiom is not a consequence of Archimedes' Axiom.** In fact in order to show with the aid of Axioms I-IV that this geometry is identical to the ordinary analytical "Cartesian" geometry Archimedes' Axiom by itself is insufficient (cf. Sections 9 and 12). However, by invoking the completeness axiom, although it contains no direct assertion about the concept of convergence, it is possible to prove the existence of a limit that corresponds to a Dedekind cut as well as the Bolzano-Weierstrass theorem for the existence of condensation points, whereby this geometry appears to be identical to Cartesian geometry.

By the above treatment the requirement of continuity has been decomposed into two essentially different parts, namely, into Archimedes' Axiom whose role is to prepare the requirement of continuity and into the completeness axiom **which forms the cornerstone of the entire system of axioms.**[1]

The subsequent investigations rest essentially only on Archimedes' Axiom and the completeness axiom is in general not assumed.

[1] Compare also the remarks at the end of Section 17, as well as my lecture on the concept of a number, "Berichte der Deutschen Mathematiker-Vereinigung, 1900." The investigation of the theorem on the equality of the two base angles of an isosceles triangle will lead to two more continuity axioms. Cf. Appendix II of this book, p. 114 and my article "Über den Satz von der Gleichheit der Basiswinkel im gleichschenkligen Dreieck," *Proceedings of the London Mathematical Society*, Vol. 35 (1903).

As additional investigations of the axioms of continuity, the following examples are mentioned here: R. Baldus, "Zur Axiomatik der Geometrie," I-III, I in *Math. Ann.*, (1928), 100, 321-33; II in *Atti d. Conge. int. d. Mat.* (Bologna, 1928), IV (1931); III in *Sitzber. d. Heidelberger Akad. Wiss.*, 1930, Fifth Proceedings. A. Schmidt, "Die Stetigkeit in der absoluten Geometrie." *Ibid.*, 1931, Fifth Proceedings. P. Bernays, "Betrachtungen über das Vollständigkeitsaxiom und verwandte Axiome," *Math. Zeitschr.* 63 (1955), 219-92.

CHAPTER II

THE CONSISTENCE AND THE MUTUAL INDEPENDENCE OF THE AXIOMS

§ 9. The Consistence of the Axioms

The axioms formulated in the five groups in Chapter I are not contradictory to each other, i.e., it is impossible to deduce from them by logical inference a result that contradicts one of them. In order to realize this a set of objects will be constructed from the real numbers in which all axioms of the five groups are satisfied.

Consider the field Ω of all algebraic numbers that arise from the number 1 and the application of a finite number of times of the four arithmetic operations of addition, subtraction, multiplication, division and the fifth operation $\left|\sqrt{1+\omega^2}\right|$, where ω denotes a number that results from these five operations.

Consider a pair of numbers (x, y) from the field Ω as a point and the ratios $(u : v : w)$ of any three numbers from Ω as a line provided u, v are not both zero. Furthermore, let the existence of the equation

$$ux + vy + w = 0$$

mean that the point (x, y) lies on the line $(u : v : w)$. Thereby, as is easy to see, Axioms I, 1-3 and IV are immediately satisfied. The numbers of the field Ω are real. Bearing in mind that these can be ordered by their magnitudes it is easy to find interpretations for these points under which all axioms of order are also valid. In fact if (x_1, y_1), (x_2, y_2), (x_3, y_3), ... are any points on a line then let this array be their sequence on this line whenever the numbers $x_1, x_2, x_3, \ldots$ or $y_1, y_2, y_3, \ldots$ in this array are monotonically decreasing or increasing. In order to satisfy further the requirements of Axiom II, 4 it is only necessary to stipulate that all points (x, y) for which $ux + vy + w$ becomes less or greater than 0 lie on one side or on the other side of the line $(u : v : w)$, respectively. It is easy to convince oneself that this

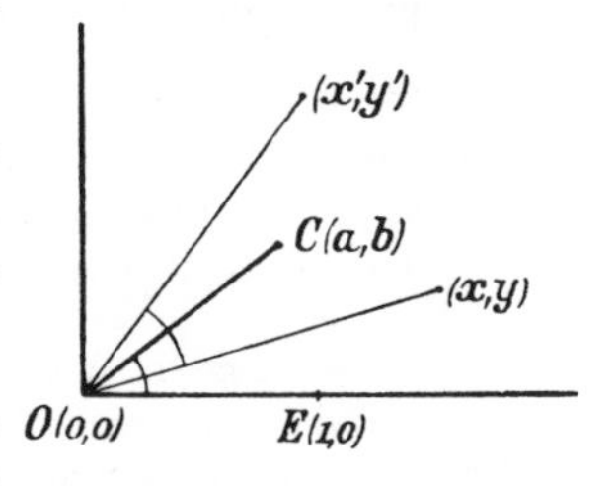

interpretation conforms to the previous one which determined the sequence of points on a line.

Construction of segments and angles follows by the well-known methods of analytic geometry. A transformation of the form

$$x' = x + a,$$
$$y' = y + b$$

produces a parallel translation of segments and angles and a transformation of the form

$$x' = x$$
$$y' = -y$$

produces a reflection about the line $y = 0$. Furthermore, denoting the point (0, 0) by O, (1, 0) by E and an arbitrary point (a, b) by C, then by a rotation through the angle $\measuredangle\, COE$, about the fixed point O the point (x, y) becomes (x', y'), where the substitutions

$$x' = \frac{a}{\sqrt{a^2+b^2}}\, x - \frac{b}{\sqrt{a^2+b^2}}\, y,$$
$$y' = \frac{b}{\sqrt{a^2+b^2}}\, x + \frac{a}{\sqrt{a^2+b^2}}\, y$$

are to be made. Since the number

$$\sqrt{a^2+b^2} = b\sqrt{1 + \left(\frac{a}{b}\right)^2}$$

is again in the field Ω Axioms III, 1-4 also hold for these interpretations and clearly the triangle congruence axioms III, 5 as well as Archimedes' Axiom V, 1 are also satisfied. The completeness axiom V, 2 is not satisfied.

Every contradiction in the consequences of the line and plane axioms I-IV, V, 1 would therefore have to be detectable in the arithmetic of the field Ω.[1]

If in the above development the field of real numbers is chosen in lieu of the field Ω the ordinary plane Cartesian geometry is obtained.

[1] Concerning the question of consistence of the arithmetic axioms, compare my lectures on the concept of a number, "Berichte der Deutschen Mathematiker-Vereinigung," 1900, as well as "Mathematische Probleme" given at the International Mathematical Congress of 1900, *Göttinger Nachrichten,* in particular, Problem 2.

The fact that in this geometry (besides Axioms I, 1-3, II, III, IV and V, 1) the axiom of completeness is also satisfied can be seen in the following way:

In Cartesian geometry it follows, on the basis of the definitions of order and segment congruence, that every segment can be divided into a given number n of congruent parts and if a segment AB is smaller than a segment AC then the n-th part of AB is also smaller than that of AC.

Assume now that there exists a line g to which, contrary to the completeness axiom, it is possible to adjoin points in the given geometry without affecting the validity of Axioms II,1-3, III, 1-3, V, 1, Theorem 5 or the uniqueness of segment construction (p. 26). Let one of these adjoined points be denoted by N. It partitions the line g into two rays each of which contains, by Archimedes' Axiom, points that existed before the extension. Designate these as the old points. N then partitions the old points of g into two rays. Considering g to be represented in parametric form

$$x = mt + n, \qquad y = pt + q$$

in which all values that the parameter t assumes before the extension by N are real then the partition induced by N yields a Dedekind cut for these values. As is well known, for such a cut only the following hold: Either the first class determined by it has a last element or the second class has a first element. Let A be the point that corresponds to one of these elements. Then no old point lies between A and N.

However, there exists an old point B such that N lies between A and B. By Archimedes' Axiom there exists a number of, say, $n-1$

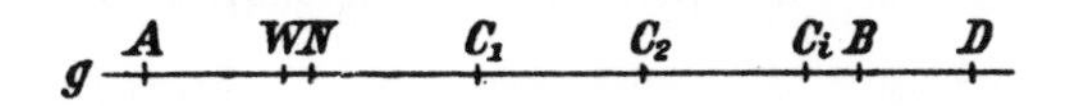

distinct points $C_1, C_2, \ldots, C_{n-2}, D$ such that the n segments AN, NC_1, C_1, $C_2, \ldots, C_{n-2}$, D are congruent to each other and such that B lies between A and D. Divide now the segment AB into n congruent parts. All dividing points are old points. Let W be the one that lies closest to A. From the requirements of line order and congruence introduced at the outset of this proof it follows that the segment AW is smaller than AN while AB is smaller than AD. The old point W lies then between A and N. The assumption that a point could be adjoined to the line g without affecting the validity of the line axioms has thus led to a contradiction.

In the above Cartesian geometry **all** line and plane axioms I-V then hold.

The corresponding treatment for space geometry presents no difficulty.

Every contradiction in the consequences of Axioms I-V would therefore be detectable in the arithmetic of the real numbers.

As can be seen, there is an infinite number of geometries which satisfy Axioms I-IV, V, 1. However, there is only **one**, namely the Cartesian geometry, in which the completeness axiom also holds at the same time.

§ 10. The Independence of the Axiom of Parallels (Non-Euclidean Geometry)[1]

Having seen the consistence of the axioms it is of interest to investigate whether they are all independent of each other. In fact it can be shown that no essential part of any one of these groups of axioms can be deduced from the others by logical inference.

As far as the individual axioms of groups I, II and III are concerned it is easy to show that the axioms of one and the same group are essentially independent of each other.

In the present exposition the axioms of groups I and II are basic to the other axioms so it is only necessary to show the independence of each of the groups III, IV and V from the others.

The axiom of parallels IV is independent of the other axioms. This is most simply shown in a well-known way as follows: Let the points, lines and planes of the ordinary (Cartesian) geometry constructed in Section 9 which lie in a fixed sphere be chosen as the elements of a space geometry and let the congruences of this geometry be replaced by linear transformations of ordinary geometry that map the fixed sphere

[1] Incidentally, it can be shown easily that in a geometry in which Axioms I-III and Archimedes' Axiom V, 1 are valid, the assertion of the axiom of parallels is either satisfied by no pair of a line a and a point A not on a, or by every such pair. Cf. R. Baldus, *Nichteuklidische Geometrie* (Berlin, 1927).

into itself. By suitable interpretations it can be seen that in this "*non-Euclidean*" *geometry* all axioms except Euclid's Axiom IV are valid and since the existence of ordinary geometry has been proved in Section 9 the existence of non-Euclidean geometry follows now.

Of special interest are the theorems that hold independently of the axiom of parallels, i.e., those which hold in Euclidean as well as in non-Euclidean geometries. As the most important examples two of Legendre's theorems will be given, the first of which requires for its proof besides Axioms I through III also Archimedes' Axiom V, 1. First some preparatory theorems will be proved:

THEOREM 33. Let a triangle OPZ with a right angle at P be given. On the segment PZ let two points X, Y be so situated that

$$\measuredangle XOY \equiv \measuredangle YOZ .$$

Then

$$XY < YZ .$$

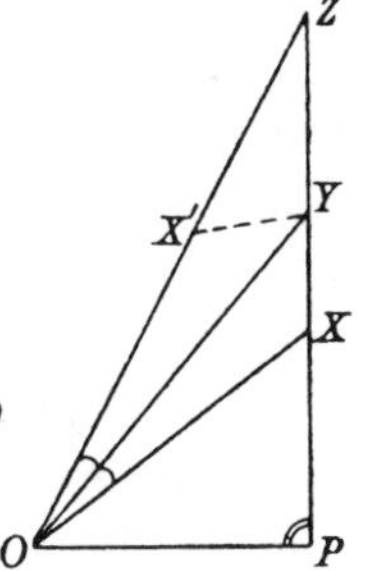

For the **proof** construct the segment OX from O on OZ so that

$$OX \equiv OX' .$$

By Theorems 22 and 23 it follows that X' lies on the segment OZ and, with the aid of Theorem 22 and Axiom III, 5, one obtains

$$\measuredangle X'ZY < \measuredangle OYX \equiv \measuredangle OYX' < \measuredangle YX'Z .$$

By Theorems 12 and 23 the relation $\measuredangle X'ZY < \measuredangle YX'Z$ leads to the assertion.

THEOREM 34. For any two angles α and ϵ it is always possible to find a natural number r such that

$$\frac{\alpha}{2^r} < \epsilon.$$

Here $\frac{\alpha}{2^r}$ denotes the angle which arises from the r-fold bisection of α.

PROOF. Let two angles α and ϵ be given. By the assumed axioms the bisection of angles is possible (see p. 23). Consider the acute angle $\frac{\alpha}{2}$. If $\frac{\alpha}{2} \leqq \varepsilon$, the assertion of Theorem 34 is true for $r = 2$. If $\frac{\alpha}{2} > \varepsilon$, then from a point C on a side at an angle $\frac{\alpha}{2}$ drop a perpendicular to the other side so that it meets it at a point B. Denote the vertex of $\frac{\alpha}{2}$ by A. If ϵ is constructed on the side AB in the interior of the angle $\measuredangle BAC = \frac{\alpha}{2}$ then by the assumed inequality the third side meets the segment BC at a point D (cf. p. 12). Archimedes' Axiom V, 1 amounts to the assertion that there exists a natural number n such that

$$n \cdot BD > BC.$$

Construct now the angle ϵ on the resulting third side towards the outside n times.

There can be a case in which at the last n-th construction, the resulting third side no longer meets the ray BC, and, say, that the m-th construction is the first one at which this occurs. Since the preceding third side still meets the side of this ray the angle $(m-1)\,\epsilon$ is acute. Hence it follows easily that the interior of the m-fold constructed angle $m\epsilon$ lies in the half plane of AB which contains C and furthermore that the ray AC lies in the interior of the angle $m\epsilon$, i.e.,

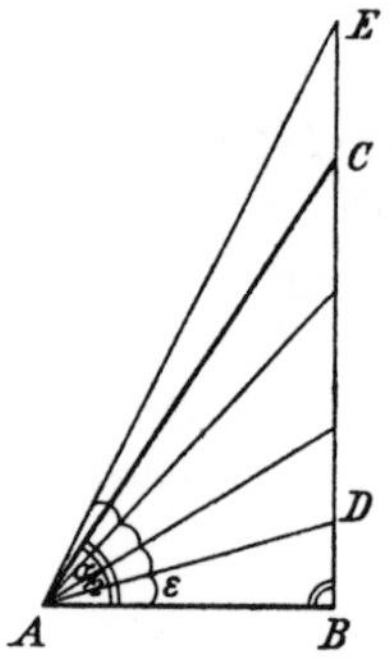

$$m \cdot \epsilon > \frac{\alpha}{2}.$$

In the other case every angle ϵ obtained in the n-fold construction delineates a segment on the ray BC which by Theorem 33 is greater or equal to BD. Let the n-th third side meet BC at the point E. The sum BE of the n delineated segments on BC is greater than $n \cdot BD$ and so a fortiori is greater than BC. Hence it follows that

$$n \cdot \epsilon > \frac{\alpha}{2}.$$

For m or n let a natural number r be determined now such that

$m < 2^{r-1}$ or $n < 2^{r-1}$, respectively. Denote the angle $m\epsilon$ or $n\epsilon$ by μ. The angles $\frac{\mu}{2^{r-1}}$ and $\frac{\alpha}{2^r}$ can be constructed. From the possibility of comparing sizes of angles one easily infers that the inequality $2^{r-1} > m$ follows from the inequality $\frac{\mu}{2^{r-1}} < \frac{\mu}{m} = \varepsilon$ that the inequality $\mu > \frac{\alpha}{2}$. Hence, by the transitivity of the quantitative comparison of (p. 21) it follows that

$$\frac{\alpha}{2^r} < \epsilon.$$

Legendre's first theorem can be proved with the aid of Theorem 34.

THEOREM 35. (**Legendre's first theorem**). The sum of the angles in a triangle is less than or equal to two right angles.

PROOF. Let any one of the three angles of a triangle be denoted by $\measuredangle A = \alpha$. Let the other two be denoted by $\measuredangle B = \beta$, $\measuredangle C = \gamma$ in such a way that $\beta \leqq \gamma$. By Theorem 26 the segment BC has a midpoint D. Extend AD beyond D by an equal amount to itself to E. By the congruence of vertical angles (p. 15) Axiom III, 5 can be applied to the triangles ADC and EDB. By defining the sum of angles on the basis of Theorem 15 in an obvious way one obtains for the angles α', β', γ' of the triangle ABE the relation

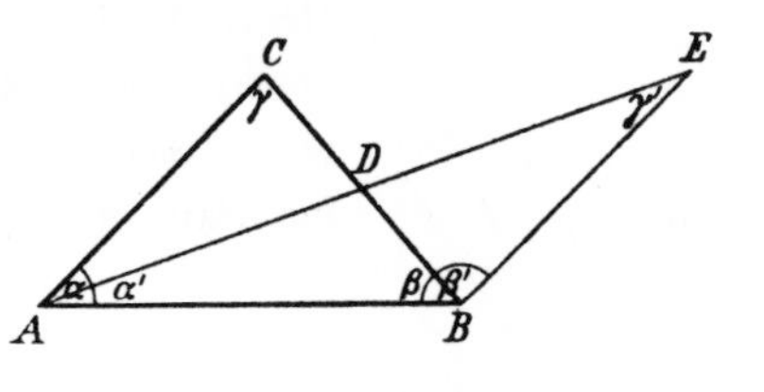

$$\alpha' + \gamma' = \alpha, \quad \beta' = \beta + \gamma.$$

The triangle ABE has thus the same sum of angles as the triangle ABC.

From the inequality $\beta \leqq \gamma$ it follows easily by Theorems 23 and 12 that

$$\alpha' \leqq \gamma' \quad \text{and hence} \quad \alpha' \leqq \frac{\alpha}{2}.$$

Thus to every triangle ABC and for any of its angles α, it is always

possible to assign a triangle with an equal sum of angles in which one angle is less than or equal to $\frac{\alpha}{2}$ and thus when a natural number r is given it is possible to assign a triangle with an equal sum of angles in which one is less than or equal to $\frac{\alpha}{2^r}$.

Assume now, contrary to the assertion of Legendre's first theorem, that the sum of the angles of the given triangle is greater than two right angles.

It follows from Theorem 22 that the sum of the angles of a triangle is less than two right triangles. Hence, according to the assumption, the sum of the angles of the given triangle can be represented in the form

$$\alpha + \beta + \gamma = 2\rho + \epsilon$$

where ϵ denotes any angles and ρ denotes a right angle. By Theorem 34 it is possible to determine a natural number r such that

$$\frac{\alpha}{2^r} < \epsilon .$$

Construct now in the specified way a triangle with the angles α^*, β^*, γ^*, satisfying the relations

$$\alpha^* + \beta^* + \gamma^* = 2\rho + \epsilon, \qquad \alpha^* \leqq \frac{\alpha}{2^r} < \epsilon.$$

In this triangle

$$\beta^* + \gamma^* > 2\rho,$$

contrary to Theorem 22. Legendre's first theorem is thus proved.

THEOREM 36. If the quadrilateral $ABCD$ has right angles at A and B and if furthermore its opposite sides AD and BC are congruent then the angles $\sphericalangle C$ and $\sphericalangle D$ are also congruent to each other. Furthermore, the perpendicular to the segment AB erected at its midpoint M meets the opposite side CD at a point N in such a way that the quadrilaterals $AMND$ and $BMNC$ are congruent.

PROOF. From Theorems 21 and 22 it follows that the

perpendicular to AB erected at M lies in the interior of the angle $\sphericalangle DMC$ and, by one of the theorems stated on p. 11, it meets the segment CD at a point N. It follows from Theorems 12, 21 and 15 that the triangles MAD and MBC and therefore also the triangles MDN and MCN, are congruent. With the aid of Theorem 15 it follows from these congruences that

$$\sphericalangle BCN \equiv \sphericalangle ADN.$$

Hence the quadrilaterals $AMND$ and $BMNC$ are congruent.

THEOREM 37. If the quadrilateral $ABCD$ has four right angles then every perpendicular EF dropped from a point E on the line CD to the opposite side AB is also perpendicular to CD.

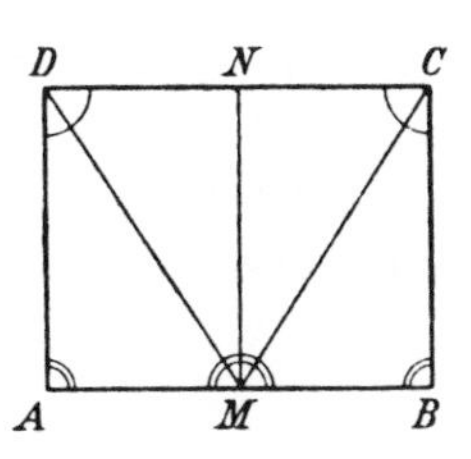

PROOF. The concept of a reflection in a line a will be introduced as follows: If a perpendicular is dropped from any point P to any line a and extended an equal amount beyond the foot to P' then the point P' is called the image of P.

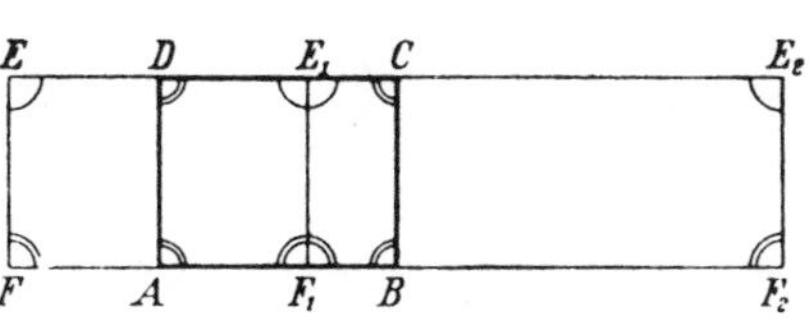

Reflect the segment EF in AD and BC. The images E_1F_1 and E_2F_2, as follows from the second part of Theorem 36, are congruent to the segment EF. The points E_1 and E_2 as well as E lie on CD: The points F_1 and F_2 as well as F lie on AB. The hypotheses of the first part of Theorem 36 are satisfied by the quadrilaterals EFF_1E_1, EFF_2E_2 and $E_1F_1F_2E_2$ and hence follows the equality of four of the angles at the points E, E_1, E_2. Therefore at one of these points two supplementary angles are equal (at E_1 in the accompanying figure), i.e., the four equal angles are right angles.

THEOREM 38. If all angles in a quadrilateral are right angles then in every quadrilateral with three right angles the fourth is also a right angle.

PROOF. Let $A'B'C'D'$ be a quadrilateral with four right angles and $ABCD$ some quadrilateral with three right angles at A, B, D. Construct the quadrilateral $AB_1C_1D_1$ that is congruent to $A'B'C'D'$ and whose right angle at A coincides with that of the quadrilateral $ABCD$.

If B coincides with B_1 or D coincides with D_1 then the assertion agrees with that of Theorem 37. If B lies between A and B_1 and D_1 lies between A and D then, as in the proof of Theorem 36, it follows from the exterior angle theorem that the segments BC and C_1D_1 intersect at a point F. Theorem 37 shows then that the angle at F, and hence also at C, is a right angle.

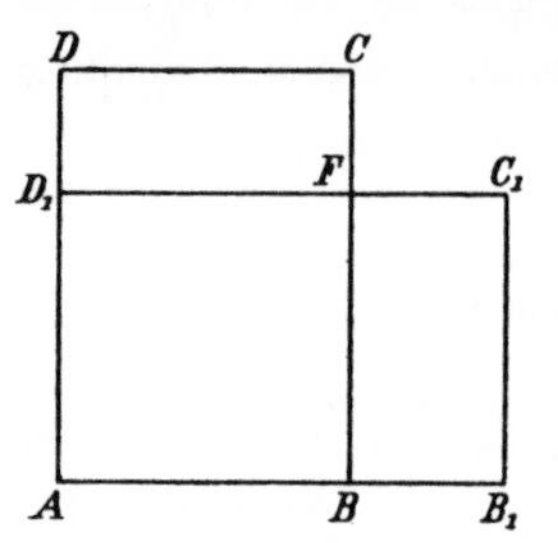

The assertions for the remaining possible orderings of the points A, B, B_1 and A, D, D_1 follow in an analogous manner:

With the aid of Theorem 28 one can prove Legendre's second theorem.

THEOREM 39. **(Legendre's second theorem).** If in some triangle the sum of the angles is equal to two right angles then the sum of the angles in every triangle is two right angles.

PROOF. With every triangle ABC whose sum of angles is $2w$ it is possible to associate a quadrilateral with three right angles whose fourth angle is equal to w. To do this join the midpoints D and E of the sides AC and BC and from A, B and C drop the perpendiculars AF, BG and CH to the connecting segment. From the congruence of the triangles AFD and CHD as well as from the congruence of the triangles BGE and CHE it follows that

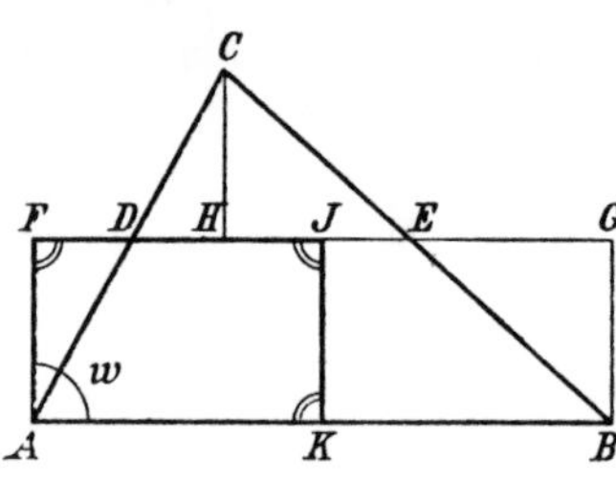

$$AF \equiv BG,$$
$$\sphericalangle FAB + \sphericalangle GBA = 2w,$$

whether one of the angles $\sphericalangle A$ and $\sphericalangle B$ of the given triangles are obtuse or not.

Erecting the perpendicular IK to FG at its midpoint it follows from the second part of Theorem 36 that the quadrilaterals $AKIF$ and $BKIG$ are congruent. Each of these quadrilaterals has three right angles and the fourth angles are equal, i.e.,

$$\sphericalangle FAB \equiv \sphericalangle GBA.$$

Hence one obtains

$$\sphericalangle FAB = w,$$

and the quadrilateral $AKIF$ is thus associated with the given triangle as desired.

Now let the sum of the angles in some triangle D_1 be equal to two right angles and let another triangle D_2 be given. Consider the associated quadrilaterals V_1 and V_2. V_1 is a quadrilateral with four right angles and V_2 is one with three right angles. By Theorem 38 the fourth angle in V_2 is also a right angle. Legendre's second theorem is thus proved.

§ 11. The Independence of the Congruence Axioms

Of the results stemming from the independence of the congruence axioms the following particularly important one will be shown—that Axiom III, 5 cannot be deduced from the other Axioms I, II, III, 1-4, IV, V by logical inference.

Choose the points, lines and planes of ordinary geometry as the elements of the new space geometry and define angle construction as in ordinary geometry, say in the manner set forth in Section 9. However, define the construction of segments in a different way. Let the two points A_1 and A_2 have the coordinates x_1, y_1, z_1 and x_2, y_2, z_2 in the ordinary geometry. Denote the length of the segment A_1A_2 by the positive value of

$$\sqrt{(x_1 - x_2 + y_1 - y_2)^2 + (y_1 - y_2)^2 + (z_1 - z_2)^2}$$

and let any two segments A_1A_2 and $A_1'A_2'$ be said to be congruent whenever they have the same length in the above sense.

It is immediately clear that in the space geometry constructed in this way Axioms I, II, III, 1-2, 4, IV, V (as well as Theorems 14, 15, 16, 19, 21 which were deduced with the aid of Axiom III, 5) hold.

In order to show that Axiom III, 3 is also satisfied choose any line a and three points A_1, A_2, A_3 on it such that A_2 lies between A_1 and A_3. Let the points x, y, z on the line a be defined by the equations

$$x = \lambda t + \lambda',$$
$$y = \mu t + \mu',$$
$$z = \nu t + \nu'$$

where t is a parameter and $\lambda, \lambda', \mu, \mu', \nu, \nu'$ denote certain constants. If $t_1, t_2\ (< t_1), t_3\ (< t_2)$ are parametric values that correspond to the points A_1, A_2, A_3 then the lengths of the three segments A_1A_2, A_2A_3 and A_1A_3 are found to be

$$(t_1 - t_2) \left| \sqrt{(\lambda + \mu)^2 + \mu^2 + \nu^2} \right|,$$
$$(t_2 - t_3) \left| \sqrt{(\lambda + \mu)^2 + \mu^2 + \nu^2} \right|,$$
$$(t_1 - t_3) \left| \sqrt{(\lambda + \mu)^2 + \mu^2 + \nu^2} \right|,$$

and the sum of the lengths of the segments A_1A_2 and A_2A_3 is thus equal to the length of the segment A_1A_2. Hence follows the validity of Axiom III, 3.

Axiom III, 5 for triangles is not always satisfied in this geometry. As an example consider the four points

O with the coordinates $x = 0, \quad y = 0$,
A " " " $x = 1, \quad y = 0$,
B " " " $x = -1, \quad y = 0$,
C " " " $x = 0, \quad y = \frac{1}{\sqrt{2}}$

in the plane $z = 0$.

The lengths of the segments OA, OB and OC are 1. For the two right triangles AOC and COB the congruences

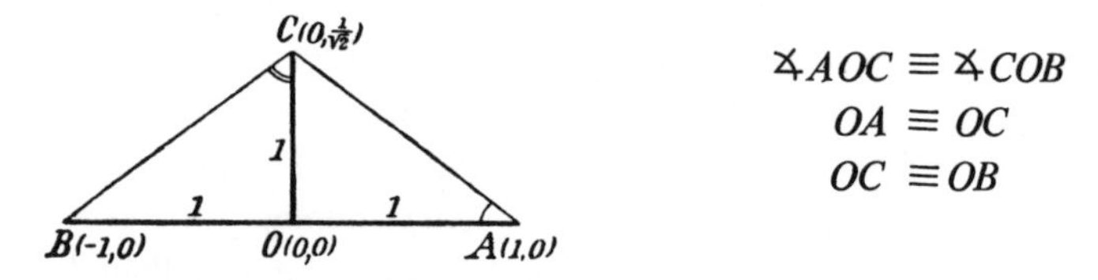

$$\measuredangle AOC \equiv \measuredangle COB$$
$$OA \equiv OC$$
$$OC \equiv OB$$

hold.

Contrary to Axiom III, 5 the angles $\measuredangle\, OAC$ and $\measuredangle\, OCB$ are not congruent. At the same time the first congruence in this example is not satisfied since the length of AC is $\sqrt{2 - \frac{2}{\sqrt{2}}}$, and that of BC is $\sqrt{2 + \frac{2}{\sqrt{2}}}$. Theorem 11 is not valid for either of the two triangles AOC or COB.

An example of a **plane** geometry in which all axioms except Axiom

III, 5 are satisfied is the following: Let all concepts that appear in the axioms with the exception of the segment congruence be defined in the usual way in a plane α. However, let the length of a segment be taken as the length of its projection, defined in the usual way, on a plane β that is inclined to α at some (non-zero) acute angle.

§ 12. The Independence of the Continuity Axiom V (Non-Archimedean Geometry)

In order to demonstrate the independence of Archimedes' Axiom V,1 it is necessary to construct a geometry in which all axioms with the exception of Axiom V are satisfied.[1]

To this end construct a field $\Omega(t)$ of all algebraic functions of t which arise from t through the five operations of addition, subtraction, multiplication, division, and the operation $\left|\sqrt{1+\omega^2}\right|$. Here ω denotes any function which is obtained by these five operations. The set of elements of $\Omega(t)$ is, as the set of those of Ω in Section 9, countable. All five operations are well defined and their results are real. Therefore the field $\Omega(t)$ contains only real and well defined functions of t.

Let c be any function of the field $\Omega(t)$. Since c is an algebraic function of t it can vanish only for a finite number of values of t; and for sufficiently large positive values of t the function c will therefore always be positive or always negative.

Regard now the functions of the field $\Omega(t)$ as some sort of complex numbers in the sense of Section 13 below. Clearly in the complex numbers defined in this manner all ordinary rules of operation are valid. Furthermore if a, b are any two distinct numbers of this set of complex numbers let the number a be greater or less than b, symbolically $a > b$ or $a < b$, according as the difference $c = a - b$ as a function of t is always positive or always negative for sufficiently large positive values of t. With this convention it is possible to order the numbers of this set of complex numbers according to their magnitude in a way that is analogous to that for the real numbers. One easily sees

[1] G. Veronese has also made an attempt to construct a geometry that is independent of Archimedes' Axiom, in his profound book, *Fondamenti di Geometria* (Padua, 1891). German translation, *Grundzüge der Geometrie*, by A. Schepp (Leipzig, 1894).

that the theorems according to which inequalities remain valid when the same number is added to both sides or when both sides are multiplied by the same positive number also hold for these complex numbers.

If n denotes any positive rational integer, then the inequality $n<t$ certainly holds for the two numbers n and t of the field $\Omega(t)$ for sufficiently large positive values of t, since the difference $n-t$, considered as a function of t, is obviously always negative. This result can be expressed as follows: The two numbers 1 and t of the field $\Omega(t)$, both of which are greater than zero, have the property that any multiple of the first remains always less than the second.

Now construct a geometry from these complex numbers of the field $\Omega(t)$ in exactly the same way that was done for the one in Section 9 based on the field Ω of algebraic numbers. Consider a set (x, y, z) of three numbers of the field $\Omega(t)$ as a point, and the ratios $(u : v : w : r)$ of any four numbers of the field $\Omega(t)$, provided u, v, w are not all zero, as a plane. Furthermore, let the existence of the equation

$$ux + vy + wz + r = 0$$

express the fact that the point (x, y, z) is in the plane $(u : v : w : r)$ and let a line be denoted by the totality of all points in two planes with different $u : v : w$. Making then suitable conventions about ordering elements and construction of segments and angles as in Section 9 a "*non-Archimedean*" *Geometry* is obtained in which, as the properties of the complex numbers $\Omega(t)$ discussed above show, all axioms with the exception of the continuity axiom are satisfied. In fact the segment 1 can be constructed contiguously on the segment t an arbitrary number of times without passing the end point of the segment t. This contradicts the requirement of Archimedes' Axiom.

The first geometry constructed in Section 9 shows that the completeness axiom V, 2 is also independent of all preceding Axioms I-IV, V, 1 since in it Archimede's Axiom is satisfied.

Both the non-Archimedean and the non-Euclidean geometries are of fundamental significance, and in particular the role that Archimedes' Axiom plays in the proof of Legendre's theorems is of great interest. The investigation of this matter which M. Dehn[1] has undertaken at my

[1] "Die Legendreschen Sätze über die Winkelsumme im Dreieck," *Math. Ann.*, Vol. 53 (1900).

urging led to a complete clarification of this problem. The investigations of M. Dehn are basic to Axioms I-III. Only at the end of Dehn's work, were axioms II of order formulated, in a more general way than in the present treatment in order also to encompass Riemannian (elliptic) geometry in the study. These can be formulated, perhaps, as follows:

Four points *A, B, C, D* on a line can always be decomposed into two pairs *A, C* and *B, D* so that *A, C* and *B, D* are "**separated**" and conversely. Five points on a line can always be labeled *A, B, C, D, E* in such a way that *A, C* are separated by *B, D* and *B, E*, that *A, D* are separated by *B, E* and *C, E*, etc.

On the basis of these Axioms I-III, and thus without the use of continuity, M. Dehn proves an extended form of Legendre's second theorem, Theorem 39.

If in *any one* triangle the sum of the angles is greater than, equal to or less than two right angles, then it is so in *every* triangle.[1]

In the above-mentioned work the following extension of Legendre's first theorem, Theorem 35, is proved.

If Archimedes' Axiom is dropped, then from the assumption of infinitely many parallels through a point it does *not* follow that the sum of the angles in a triangle is less than two right angles. Moreover, there exists a geometry (the non-Legendrian geometry) in which it is possible to draw through a point infinitely many parallels to a line and in which nevertheléss the theorems of Riemannian (elliptic) geometry hold. On the other hand there exists a geometry (the semi-Euclidean geometry) in which there exists infinitely many parallels to a line through a point and in which the theorems of Euclidean geometry still hold.

From the assumption that there exist no parallels it always follows that the sum of the angles in a triangle is greater than two right angles.

Finally, it can be observed that if **Archimedes' Axiom is invoked** then the axiom of parallels can be replaced by the requirement that the sum of the angles in a triangle be two right angles.

[1] A proof of this theorem was also produced later by F. Schur, *Math. Ann.*, Vol. 55, and again by Hjelmslev, *Math. Ann.*, Vol. 64. In the latter, the extremely short argument that leads to the proof of the middle part of this theorem is noteworthy. Cf. also F. Schur, *Grundlagen der Geometrie* (Leipzig and Berlin, 1909), Section 6.

CHAPTER III

THEORY OF PROPORTION

§ 13. Complex Number Sets[1]

This chapter will be prefaced by a brief discussion of complex numbers which will be particularly useful later on for clarifying the exposition.

The real numbers as a whole form a set of objects with the following properties:

THEOREMS OF COMPOSITION (1-6):

1. The number a and the number b generate by "**addition**" a definite number c. Symbolically

$$a + b = c \qquad \text{or} \qquad c = a + b.$$

2. If a and b are given numbers then there always exists one and only one number x and also one and only one number y such that

$$a + x = b \qquad \text{and} \qquad y + a = b.$$

3. There exists a definite number, denoted by 0, such that for every a both

$$a + 0 = a \qquad \text{and} \qquad 0 + a = a.$$

4. The number a and the number b generate in another way, by "**multiplication**," a definite number c. Symbolically

$$ab = c \qquad \text{or} \qquad c = ab.$$

5. If a and b are any given numbers, and a is not 0 then there always exists one and only one number x, and also one and only one number y, such that

$$ax = b \qquad \text{and} \qquad ya = b.$$

6. There exists a definite number, denoted by 1, such that for every a both

$$a \cdot 1 = a \qquad \text{and} \qquad 1 \cdot a = a.$$

[1] See also Supplement I, 2.

RULES OF OPERATION (7-12):

If a,b,c are any numbers then the following rules of operation hold:

7. $$a + (b + c) = (a + b) + c$$
8. $$a + b = b + a$$
9. $$a(bc) = (ab)c$$
10. $$a(b + c) = ab + ac$$
11. $$(a + b)c = ac + bc$$
12. $$ab = ba.$$

THEOREMS OF ORDER (13-16):

13. If a,b are any two distinct numbers then one and only one of them (say a) is always greater than the other. The latter is then called the smaller number. Symbolically

$$a > b \qquad \text{and} \qquad b < a.$$

$a > a$ holds for no number a.

14. If $a > b$ and $b > c$, then $a > c$.

15. If $a > b$, then

$$a + c > b + c.$$

16. If $a > b$ and $c > 0$, then

$$ac > bc.$$

THEOREMS OF CONTINUITY (17-18):

17. **(Archimedes' Theorem).** If $a > 0$ and $b > 0$ are any two numbers, then it is always possible to add a to itself a sufficient number of times so that the resulting sum is greater than b. Symbolically

$$a + a + \cdots + a > b.$$

18. **(Theorem of Completeness).** It is impossible to adjoin to the set of numbers another set of objects, as numbers, so that Theorems 1-17 are also satisfied in the set arising from the adjuction when the relations among the numbers are preserved; or briefly, the numbers form a set of objects which, by preserving all relations and formulated theorems, admits no extension.

Let a set of objects which has only some of the Properties 1-18 be called a set of *complex numbers*. A set of complex numbers will be called *Archimedean* or *non-Archimedean* according as it does or does not satisfy Requirement 17.

Some of the formulated Properties 1-18 **are consequences of the others.** The problem now is to investigate the logical dependence of these properties.[1] Because of their geometric significance two questions of a similar nature will be answered in Chapter VI, Sections 32 and 33. Here it will only be shown that Requirement 17 is by no means a logical consequence of the preceding properties, since as an example, the set of complex numbers Ω (t) considered in Section 12 has all Properties 1-16, but does not satisfy Requirement 17.

Otherwise, the corresponding remarks made in Section 8 about the geometric axioms of continuity are valid for the Theorems of Continuity (17-18).

§ 14. Proof of Pascal's Theorem

In this and in the following chapter the **plane** axioms of all groups with the exception of the continuity axiom, i.e., Axioms I, 1-3 and II-IV, are assumed for the investigation. In the present chapter it is intended to establish Euclid's theory of proportion by means of the above-mentioned axioms, i.e., *in the plane and independently of the Archimedean Axiom* (see also Supplement II).

For this purpose, a result that is a special case of Pascal's well-known theorem in the theory of conic sections will be proved next.

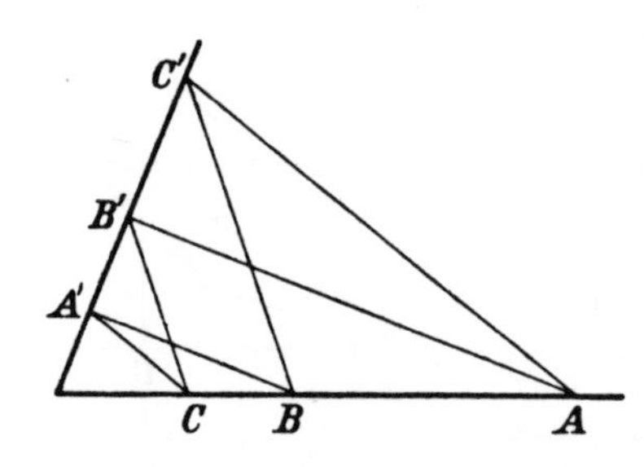

This theorem, which will be referred to hereinafter as Pascal's Theorem, is stated as follows:

THEOREM 40[2] **(Pascal's Theorem).**

Let A, B, C, and A', B', C', be two sets of points on two intersecting lines that are distinct from the point of intersection of the lines. If CB' is parallel to BC' and CA' is parallel to AC' then BA' is also parallel to AB'.

[1] Cf. Supplement I, 2.

[2] F. Schur has published in *Math. Ann.*, Vol. 51, an interesting proof of Pascal's Theorem based on the plane and space axioms I-III. So did Dehn in *Math. Ann.*, Vol. 53. Later, J. Hjelmslev, using the results of G. Hessenberg (*Math. Ann.*, Vol. 61) succeeded in proving Pascal's Theorem on the basis of only the plane axioms I-III ("Neue Begründung der ebenen Geometrie," *Math. Ann.*, Vol. 64). Cf. Appendix III of this book.

For the proof of this theorem the following notation is introduced: In a right triangle the leg a is clearly uniquely determined by the hypotenuse c and the base angle α included between a and c. Briefly stated

$$a = \alpha c,$$

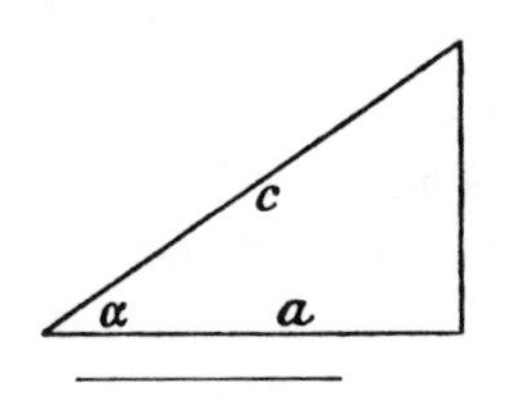

so that the symbol αc always denotes a definite segment whenever c is any given segment and α is any given acute angle. In the same way a segment c is uniquely determined by any given segment a and by any given acute angle α through the equation

$$a = \alpha c.$$

Now let c denote any segment and α, β any two acute angles. It will be observed that the segment congruence

$$\alpha\beta c \equiv \beta\alpha c$$

always holds and thereby the symbols α, β can always be interchanged.

To prove the assertion, construct, on the segment $c = AB$, with A as a vertex, the angle α on one side and the angle β on the other side. Then drop from B the perpendiculars BC and BD to the other sides of these angles, join C with D and finally drop from A the perpendicular AE to CD.

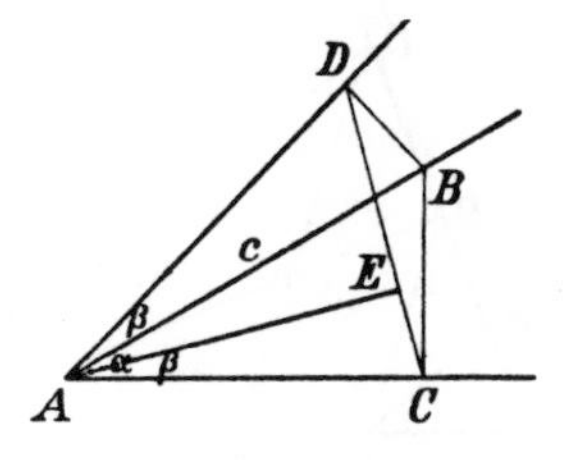

Since the angles $\measuredangle\ ACB$ and $\measuredangle\ ADB$ are right angles the four points A, B, C, D, lie on a circle and hence the two angles $\measuredangle\ ACD$ and $\measuredangle\ ABD$, as inscribed angles on the same chord AD, are congruent. Now $\measuredangle\ ACD$ together with $\measuredangle\ CAE$, and $\measuredangle\ ABD$ together with $\measuredangle\ BAD$ constitute right angles and consequently the angles $\measuredangle CAE$ and $\measuredangle\ BAD$ are also congruent, i.e.,

$$\measuredangle\ CAE \equiv \beta$$

and hence

$$\measuredangle\ DAE \equiv \alpha.$$

The segment congruences

$\beta c \equiv AD,$	$\alpha c \equiv AC,$
$\alpha\beta c \equiv \alpha(AD) \equiv AE,$	$\beta\alpha c \equiv \beta(AC) \equiv AE,$

are obtained immediately and hence follows the validity of the aforementioned congruence.

Returning now to the figure of Pascal's Theorem, denote the point of intersection of the two lines by O and the segments $OA, OB, OC, OA', OB', OC', CB', BC', AC', CA', BA', AB'$ by $a, b, c, a', b', c', l, l^*, m, m^*, n, n^*$, respectively. Then drop perpendiculars from O to l, m^*, n. Let the perpendicular to l form with the lines OA, OA' the acute angles λ', λ and let the perpendiculars to m^* and to n form the angles μ', μ and ν', ν, respectively. Expressing now these three perpendiculars in the above indicated manner in terms of the hypotenuses and the base angles of the formed right triangles in two ways one obtains the following three congruences:

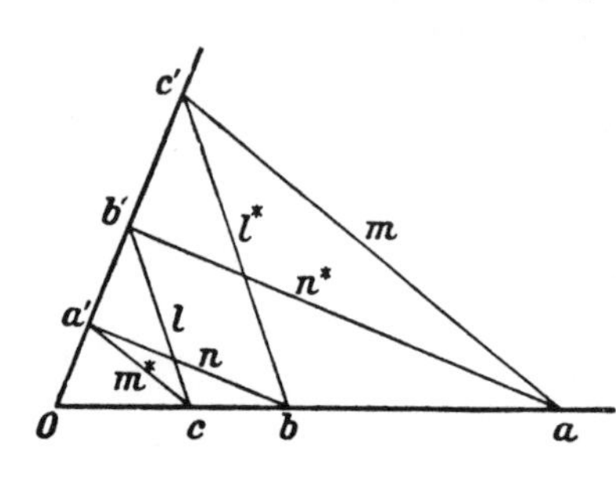

(1) $$\lambda b' \equiv \lambda' c,$$

(2) $$\mu a' \equiv \mu' c,$$

(3) $$\nu a' \equiv \nu' b.$$

Since by hypothesis l is parallel to l^* and m is parallel to m^*, the perpendiculars from O to l^* and to m must coincide with those to l and m^*, respectively, and thereby one obtains

(4) $$\lambda c' \equiv \lambda' b,$$

(5) $$\mu c' \equiv \mu' a.$$

If the symbol $\lambda'\mu$ is applied to the left- and right-hand sides of congruence (3) and it is recalled that according to what has been shown before these symbols are interchangeable, it is found that

$$\nu\lambda'\mu a' \equiv \nu'\mu\lambda' b.$$

In this congruence substitute congruence (2) on the left and congruence (4) on the right. Then

$$\nu\lambda'\mu' c \equiv \nu'\mu\lambda c'$$

or

$$\nu\mu'\lambda' c \equiv \nu'\lambda\mu c'.$$

Substituting herein congruence (1) on the left and (5) on the right

$$\nu\mu'\lambda b' \equiv \nu'\lambda\mu' a$$

or

$$\lambda\mu'\nu b' \equiv \lambda\mu'\nu' a.$$

On the basis of one of the properties of the symbols given on p. 47, one concludes immediately from the last congruence that

$$\mu'\nu b' \equiv \mu'\nu' a$$

and hence

$$\nu b' \equiv \nu' a. \tag{6}$$

Considering now the perpendicular to n dropped from O and dropping perpendiculars to it from A and B', congruence (6) shows that the feet of the last two perpendiculars coincide, i.e., the line $n^* = AB'$ is perpendicular to the perpendicular to n and is therefore parallel to n. The proof of Pascal's Theorem is thus complete.

In order to establish the theory of proportion only the special case of Pascal's Theorem will be used in which the segment congruence

$$OC \equiv OA'$$

and hence also

$$OA \equiv OC'$$

holds and in which the points A, B, C lie on the same ray emanating from O. In this special case the proof is particularly easy, namely as follows:

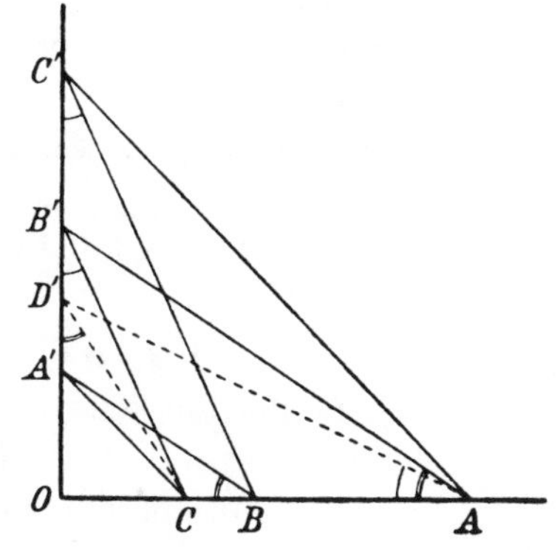

From O lay off the segment OB on OA' to D' so that the connecting line BD' is parallel to CA' and AC'. In view of the congruence of the triangles $OC'B$ and OAD'

(1†) $\quad \measuredangle\, OC'B \equiv \measuredangle\, OAD'.$

Since by hypothesis CB' and BC' are parallel

(2†) $\quad \measuredangle\, OC'B \equiv \measuredangle\, OB'C;$

From (1†) and (2†) one deduces

$$\measuredangle\, OAD' \equiv \measuredangle\, OB'C;$$

Since by a property of circles $ACD'B'$ is an inscribed quadrilateral the congruence

(3†) $$\measuredangle\, OD'C \equiv \measuredangle\, OAB'$$

follows from a well-known theorem about the angles of an inscribed quadrilateral. On the other hand, in view of the congruence of the triangles $OD'C$ and OBA'

(4†) $$\measuredangle\, OD'C \equiv \measuredangle\, OBA'.$$

From (3†) and (4†) one deduces that

$$\measuredangle\, OAB' \equiv \measuredangle\, OBA',$$

and this congruence shows that AB' and BA' are parallel to each other, as required by Pascal's Theorem.

Given any line, a point not on it and any angle, it is clear that by constructing this angle and by drawing a parallel it is possible to find a line that passes through the given point and intersects the given line at the given angle. In view of this it is even possible to use for the proof of Pascal's more general theorem the following simple argument, for which I am indebted to another source.

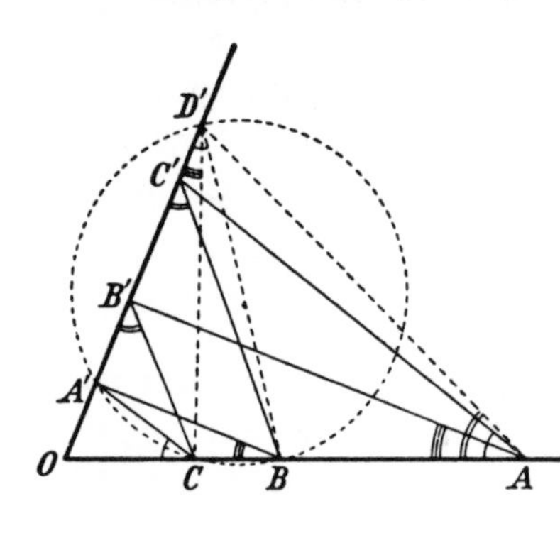

Through B draw a line that meets OA' at the point D' at the angle $\measuredangle\, OCA'$ so that the congruence

(1*) $$\measuredangle\, OCA' \equiv \measuredangle\, OD'B$$

holds. Then by a well-known theorem on circles $CBD'A'$ is an inscribed quadrilateral, and by the theorem of congruence of inscribed angles on the same chord follows the congruence

(2*) $$\measuredangle\, OBA' \equiv \measuredangle\, OD'C.$$

Since by hypothesis CA' and AC' are parallel

(3*) $$\measuredangle\, OCA' \equiv \measuredangle\, OAC'.$$

From (1*) and (3*) one deduces the congruence

$$\measuredangle\, OD'B \equiv \measuredangle\, OAC'.$$

Then however $BAD'C'$ is also an inscribed quadrilateral and by the quadrilateral angle theorem the congruence

(4*) $$\measuredangle\, OAD' \equiv \measuredangle\, OC'B$$

holds. Furthermore, since by hypothesis CB' is parallel to BC' one also has

(5*) $$\measuredangle\, OB'C \equiv \measuredangle\, OC'B.$$

From (4*) and (5*) one deduces the congruence

$$\measuredangle\, OAD' \equiv \measuredangle\, OB'C.$$

This one finally shows that $CAD'B'$ is an inscribed quadrilateral and hence the congruence

(6*) $$\measuredangle\, OAB' \equiv \measuredangle\, OD'C$$

also holds.

From (2*) and (6*) follows

$$\measuredangle\, OBA' \equiv \measuredangle\, OAB',$$

and this congruence shows that BA' and AB' are parallel as required by Pascal's Theorem.

If D' coincides with one of the points A', B', C' or if the order of the points A, B, C is different a change in this argument is necessary, as is easy to see.[1]

§ 15. The Arithmetic of Segments Based on Pascal's Theorem

Pascal's Theorem proved in the preceding paragraphs prepares the ground for the introduction into geometry of an arithmetic of segments in which all the rules of operation of the real numbers hold with no change.

In the arithmetic of segments, the word "equal" will be used instead of "congruent" and the sign "=" instead of "≡".

If A, B, C are three points on a line and B lies between A and C the *sum* of the two segments $a = AB$ and $b = BC$ is denoted by $c = AC$ and is expressed as

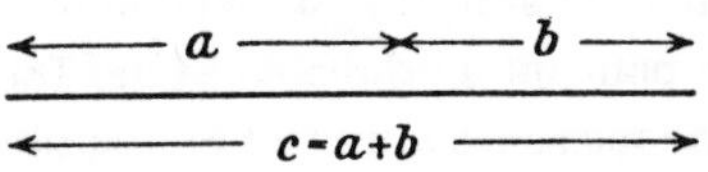

a b

$c = a+b$

$$c = a + b.$$

The segments a and b are said to be smaller than c. Symbolically,

$$a < c, \qquad b < c,$$

and c is said to be larger than a and b. Symbolically,

$$c > a, \qquad c > b.$$

[1] The use which the theorem on the point of intersection of the altitudes of a triangle finds in the establishment of Pascal's Theorem or in the theory of proportion also deserves some interest. About this cf. F. Schur, *Math. Ann.*, Vol. 57, and J. Mollerup, *Studier over den plane geometris Aksiomer* (Copenhagen, 1903).

From the line congruences III, 1-3 it is easy to infer that for the addition of segments thus defined, the **associative** law

$$a + (b + c) = (a + b) + c$$

as well as the **commutative** law $a + b = b + a$ is valid.

In order to define geometrically the product of a segment a by a segment b the following construction will be used: Choose any segment which remains fixed during the entire discussion and denote it by 1. Now lay off the segments 1 and b from the vertex O on a side of a right triangle. Then lay off the segment a on the other side. Join the end points of the segments 1 and a with a line and through the end point of the segment b draw a parallel to this line. It will delineate a segment c on the other side. This segment is then called the *product* of the segment a by the segment b and is denoted by

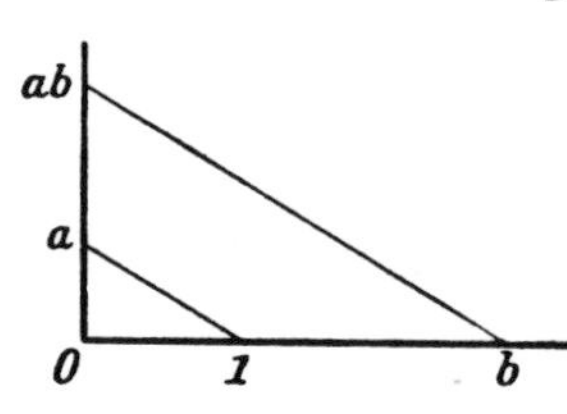

$$c = ab.$$

It will be shown first that the multiplication of segments thus defined follows the **commutative** law

$$ab = ba.$$

To do this construct in the manner prescribed above the segment ab. Then lay off the segment a on the first side of the right angle and the segment b on the other side, join by a line the end point of the segment 1 with the end point of the segment b on the other side and draw a parallel to this line through the end point of a on the first side. The latter delineates on the other side the segment ba. In fact by Pascal's Theorem (Theorem 40), in view of the parallelism of the auxiliary dotted lines shown in the figure, the segment ba coincides with the previously constructed segment ab. Conversely it also follows from the validity of the commutative law in the segment arithmetic, as one sees immediately, that the special case of Pascal's Theorem referred to on p. 46 also holds for figures in which the rays OA and OA' form a right angle.

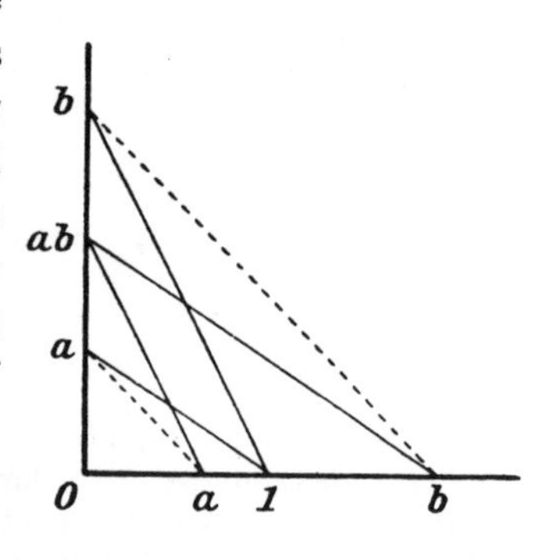

In order to prove the **associative** law

$$a(bc) = (ab)c$$

for segment multiplication, construct from the vertex O the segments 1 and b on one side of the right angle and also from O the segments a and c on the other side. Then construct the segments $d = ab$ and $e = cb$ and from O lay off these segments on the first side. Constructing then ae and cd it follows again by Pascal's Theorem, as is apparent from the accompanying figure, that the end points of these segments coincide, i.e.,

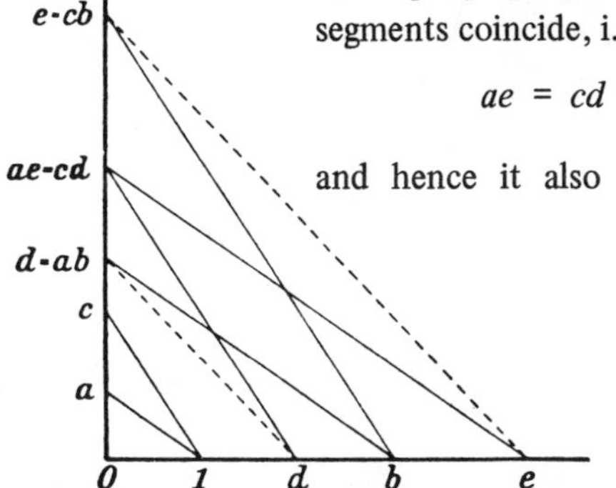

$$ae = cd \quad \text{or} \quad a(cb) = c(ab),$$

and hence it also follows with the aid of the commutative law that

$$a(bc) = (ab)c.$$[1]

As can be seen in the above proof of the commutative as well as of the associative laws of multiplication use was made only of the special case of Pascal's Theorem whose proof on pp. 49 to 50 (Section 14) can be carried out in a particularly simple way by several applications of the inscribed quadrilateral theorem.

By combining these developments the following method for multiplication of segments is arrived at, which among all methods encountered so far seems to be the simplest.

[1] Compare to this, also, the methods of the development of the theory of proportion which have been given in the meantime by A. Kneser, *Archiv für Math. und Phys.*, Series III, Vol. 2, and by J. Mollerup, *Math. Ann.*, Vol. 56, as well as *Studier over den plane geometris Aksiomer* (Copenhagen, 1903), in which the proportion equation is assumed. In "Zur Proportionenlehre," F. Schur remarks that Kupffer (*Sitzungsber. der Naturforschergesellschaft zu Dorpat*, 1893), has already correctly proved the commutative law of multiplication. However, the rest of Kupffer's development of the theory of proportion should be regarded as inadequate.

Lay off the segments $a = OA$ and $b = OB$ from the vertex O on one of the sides of a right angle and on the other side the unit segment $1 = OC$. Let the circle through the points A, B, C intersect the latter side at the point D. The point D can easily be obtained by the congruence axiom alone, and without the use of compasses, by dropping from the center of the circle a perpendicular to OC and reflecting in it the point C. In view of the equality of the angles $\sphericalangle\ OCA$ and $\sphericalangle\ OBD$ it follows by the definition of the product of two segments (p. 52) that

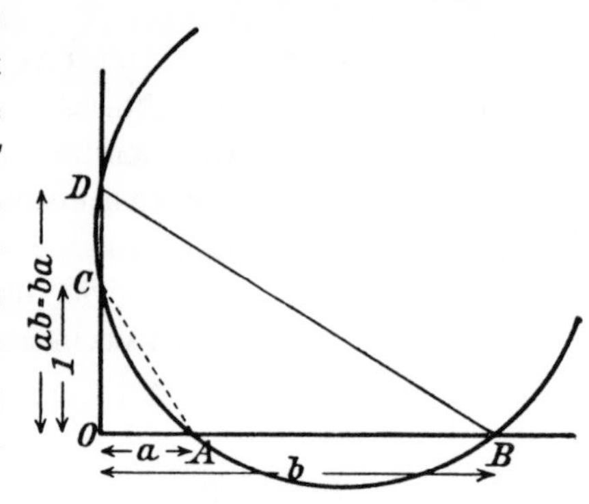

$$OD = ab,$$

and in view of the equality of the angles $\sphericalangle\ ODA$ and $\sphericalangle\ OBC$ it follows by the same definition that

$$OD = ba.$$

By a remark on p. 52 the commutative law of multiplication

$$ab = ba$$

that follows from this *shows* now that the special case of Pascal's Theorem referred to on p. 49 holds for the sides of a right angle, and from this in turn, by p. 52, follows the associative law of multiplication

$$a(bc) = (ab)c.$$

Finally in this arithmetic of segments the **distributive** law

$$a(b + c) = ab + ac$$

also holds.

In order to prove this, construct the segments ab, ac and $a(b + c)$ and through the end points of the segment c (see accompanying figure) draw a parallel to the other side of the right angle. The congruence of the two right angles, the shaded triangles in the figure, and an application of the theorem on the equality of opposite sides in a parallelogram yield the desired proof.

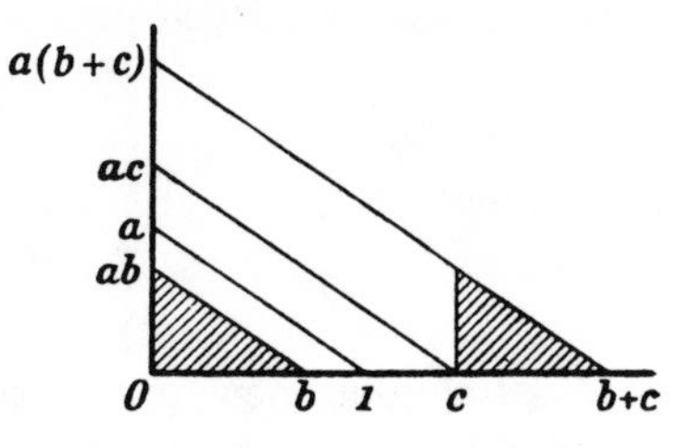

If b and c are any two segments then there always exists a segment a such that $c = ab$. The segment a will be denoted by $\frac{c}{b}$ and called the *quotient* of c and b.

§ 16. Proportion and the Similarity Theorems

With the aid of the segment arithmetic established above, Euclid's theory of proportion can satisfactorily be developed in the following manner without the use of Archimedes' Axiom:

DEFINITION. If a, b, a', b' are any four segments let the *proportion*

$$a : b = a' : b'$$

denote nothing else but the segment equation

$$ab' = ba'.$$

DEFINITION. Two triangles are called *similar* if their corresponding angles are congruent.

THEOREM 41. If a, b and a', b' are corresponding sides of two triangles then the proportion

$$a : b = a' : b'$$

holds.

PROOF. Consider the special case in which the angles in both triangles included by the sides a, b and a', b' are right angles and assume that both triangles are inscribed in one and the same right angle. Then lay off from the vertex the segment 1 on one side and through the end point of this segment draw a parallel to both hypotenuses. Let the latter delineate on the other side the segment e. Then by the definition of segment multiplication

$$b = ea, \qquad b' = ea';$$

and hence one obtains

$$ab' = ba',$$

i.e.,

$$a : b = a' : b'.$$

Returning now to the general case, in each of the two similar triangles construct the points of intersection of the angle bisectors S and S', whose existence is easy to deduce from Theorem 25, and from them drop the perpendiculars r and r' to the sides of the triangles.

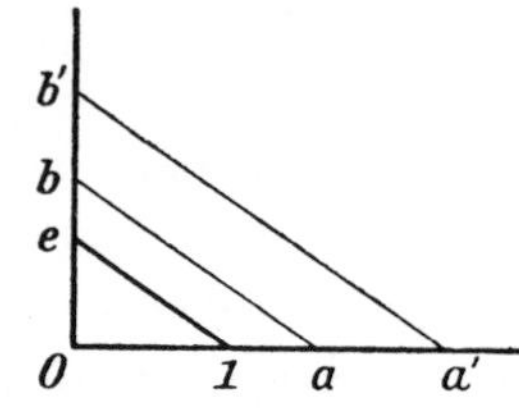

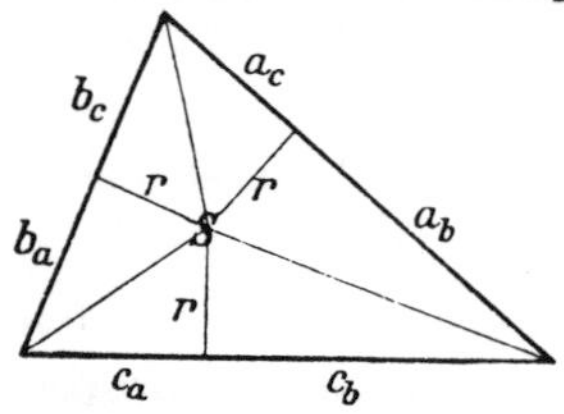

Denote the segments delineated in this way by

$$a_b,\ a_c,\ b_c,\ b_a,\ c_a,\ c_b$$

and

$$a'_b, a'_c,\ b'_c,\ b'_a,\ c'_a,\ c'_b.$$

The special case of this theorem proved before yields the proportions

$$\begin{array}{r|r} a_b : r = a'_b : r' & b_c : r = b'_c : r' \\ a_c : r = a'_c : r' & b_a : r = b'_a : r'; \end{array}$$

from which, with the aid of the distributive law, one deduces

$$a : r = a' : r', \quad b : r = b' : r'$$

and hence

$$b'ar' = b'ra', \quad a'br' = a'rb'.$$

With the aid of the commutative law of multiplication, these equations yield

$$a : b = a' : b'.$$

From Theorem 41 one deduces easily the fundamental theorem of the theory of proportion which reads as follows:

THEOREM 42. *If two parallels delineate on the sides of any angle the segments a, b, and a', b' then the proportion*

$$a : b = a' : b'$$

holds.

Conversely, if four segments a, b, a′, b′ satisfy this proportion and a, a′, and b, b′ are constructed in pairs on the sides of any angle then the lines joining the end points of a, b and a′, b′ are parallel.

§ 17. The Equations of Lines and Planes

To the hitherto considered set of segments adjoin another set of such segments. By the axioms of order it is easy to distinguish on a line a "*positive*" and a "*negative*" direction. A segment AB that has been denoted thus far by a will continue to be denoted by a only if B lies in the positive direction of A; otherwise it will be denoted by $-a$. A point will be denoted as the segment 0. The segment a is said to be "*positive*" or greater than 0, symbolically $a > 0$; the segment $-a$ is said to be "*negative*" or less than 0, symbolically $-a < 0$.

In this extended arithmetic of segments all Rules of Operation 1-16, which were listed in Section 13, are valid. The following special results are emphasized:

$$a \cdot 1 = 1 \cdot a = a \quad \text{and} \quad a \cdot 0 = 0 \cdot a = 0$$

always hold. If $ab = 0$ then either $a = 0$ or $b = 0$. If $a > b$ and $c > 0$ then it always follows that $ac > bc$. Furthermore if $A_1, A_2, A_3, \ldots, A_{n-1}, A_n$ are any n points on a line then the sum of the segments $A_1A_2, A_2A_3, \ldots, A_{n-1}A_n, A_nA_1$ is equal to zero.

In a plane α assume now two mutually perpendicular lines through the point O as fixed coordinate axes and lay off from O any segments x, y on the two lines. Then erect perpendiculars to these lines at the end points of the segments x, y and determine their point of intersection P. The segments x, y are then called the coordinates of the point P. Every point in the plane α is uniquely determined by its coordinates x, y which can be positive or negative segments or 0.

Let l be any line in the plane α that passes through the points O and C with coordinates a, b. If x, y are the coordinates of any point on l then it is easily found by Theorem 42 that

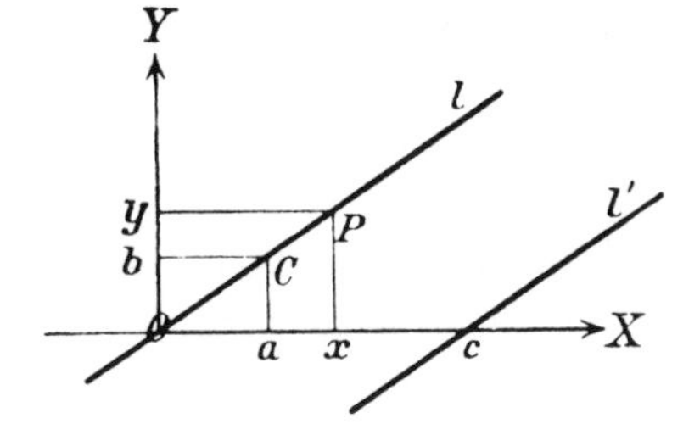

$$a : b = x : y$$

or that

$$bx - ay = 0$$

is the equation of the line l. If l' is a line parallel to l that delineates the segment c on the x-axis then the equation of the line l' can be obtained by substituting the segment $x - c$ for the segment x in the equation for the line l. The desired equation then becomes

$$bx - ay - bc = 0.$$

From these developments it is easy to conclude in a way that is independent of the Archimedean Axiom that every line in a plane can be represented by means of a linear equation by the coordinates x, y and conversely, every such linear equation, in which the coefficients are segments in the given geometry, represents a line.

The corresponding results in space geometry can be shown just as easily.

The further development of geometry can be done from now on by the ordinary methods that are used in analytic geometry.

So far the Archimedean Axiom has not been employed in this chapter. If its validity is assumed now then it is possible to assign real numbers to the points of any line in space in the following manner:

Choose any two points on a line and assign to them the numbers 0 and 1. Then bisect the segment 01 determined by them and denote the resulting midpoint by $\frac{1}{2}$. Then denote the midpoint of the segment $0\frac{1}{2}$ by $\frac{1}{4}$, etc. After carrying out n times this scheme a point is reached which is to be assigned the number $\frac{1}{2^n}$. Then lay off the segment $0\frac{1}{2^n}$ m times contiguously from the point 0 towards the point 1 as well as to the other side and assign to the resulting points the numbers $\frac{m}{2^n}$ and $-\frac{m}{2^n}$, respectively. On the basis of this assignment it can easily be deduced by the Archimedean Axiom that any point on a line can be assigned as real number in a uniquely determined way, and indeed, that this assignment has the following property: If A, B, C are any three points on a line and α, β, γ are their corresponding real numbers and B lies between A and C then these numbers always satisfy the inequality

$$\alpha < \beta < \gamma \quad \text{or} \quad \alpha > \beta > \gamma.$$

From the developments in Section 9 of Chapter II, it is evident that for every number of the algebraic field Ω there must exist a point on the line to which it can be assigned. Whether there correspoinds a point to any other real number depends on whether the completeness axiom V, 2 holds in the given geometry.

However, if in a geometry only the validity of the Archimedean Axiom is assumed then it is possible to extend the set of points, lines and planes by "**irrational**" elements so that in the resulting geometry on every line there corresponds a point without exception to every set of three real numbers that satisfy its equation. By suitable interpretations it is possible to infer at the same time that **all** Axioms I-V are valid in the extended geometry. This extended geometry (by the adjunction of irrational elements) is none other than the ordinary space Cartesian geometry in which the completeness axiom V, 2 also holds.[1]

[1] Cf. the remarks at the end of Section 8.

CHAPTER IV

THEORY OF PLANE AREA

§ 18. Equidecomposability and Equicomplementability of Polygons

For the investigations of this chapter the same axioms are assumed that were laid down for the third chapter, namely, the line and plane axioms of all groups with the exception of the continuity axiom, i.e., Axioms I, 1-3, and II-IV.

The theory of proportion discussed in Chapter III and the segment arithmetic introduced there make it possible to develop Euclid's theory of area with the aid of the aforementioned axioms, i.e., to develop it *in the plane independently of the axiom of continuity*.

Since by the development in Chapter II the theory of proportion rests essentially on Pascal's Theorem (Theorem 40) the same is true of the theory of area. This development of the theory of area appears as one of the most remarkable applications of Pascal's Theorem in elementary geometry.

DEFINITION. If two points of a simple polygon P are joined by some polygonal segment that lies entirely in the interior of the polygon and which has no double point, two new simple polygons P_1 and P_2 are formed whose interior points lie in the interior of P. P is then said to **decompose** into P_1 and P_2 or P **is decomposed into** P_1 and P_2 or P_1 and P_2 compose P.

DEFINITION. Two simple polygons are called *equidecomposable* if they can be decomposed into a finite number of triangles that are congruent in pairs.

DEFINITION. Two simple polygons P and Q are called *equicomplementable* if it is possible to adjoin to them a finite number of pairs of equidecomposable polygons P', Q'; P'', Q''; ...; P''', Q''' such that the composed polygons $P + P' + P'' + \ldots + P'''$ and $Q + Q' + Q'' + \ldots + Q'''$ are equidecomposable with each other.

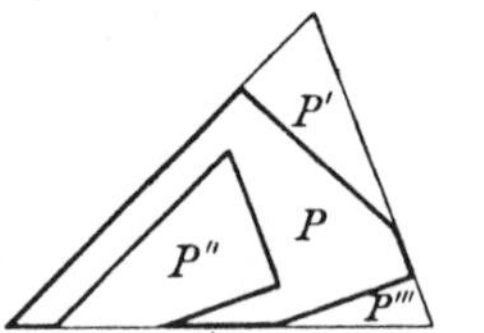

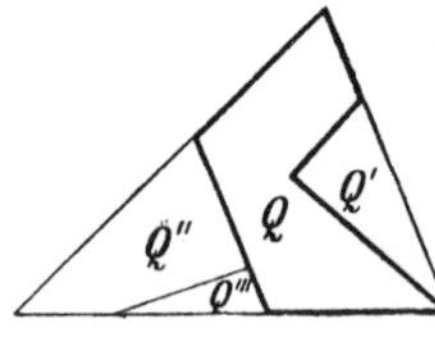

From these definitions it follows immediately that combining equidecomposable polygons results in **equidecomposable** polygons again and if equidecomposable polygons are removed from equidecomposable polygons the remaining polygons are **equicomplementable.**

Furthermore the following theorems hold:

THEOREM 43. If two polygons P_1 and P_2 are equidecomposable

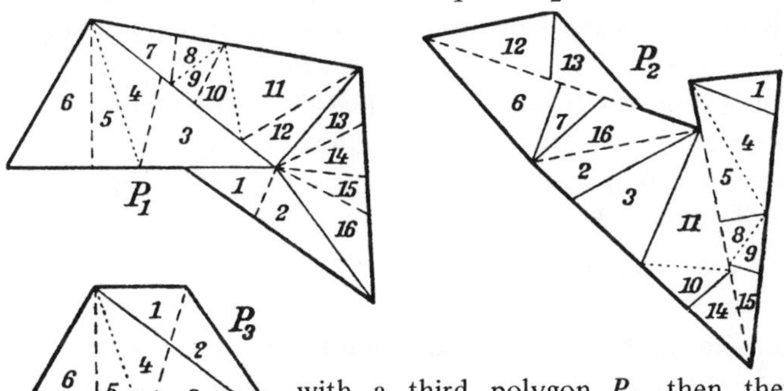

with a third polygon P_3 then they are equidecomposable with each other. If two polygons are equicomplementable with a third one then they are equicomplementable with each other.

PROOF. By hypothesis a decomposition into triangles is possible for P_1 as well as for P_2 so that to each of these two decompositions there corresponds a decomposition of P_3 into congruent triangles. If both of these decompositions of P_3 are considered simultaneously then generally every triangle of one decomposition will be decomposed into polygons by segments of the other decomposition. Now add a sufficient number of segments so that every one of these polygons itself splits into triangles and apply the two resulting decompositions into triangles to P_1 and P_2. Then evidently the two polygons P_1 and P_2 split into an equal number of congruent triangles in pairs and thus are, by definition, equidecomposable with each other.

The second assertion of Theorem 43 follows now with no difficulty (see Supplement III). The concepts of *rectangle, base* and *altitude of a parallelogram, base* and *altitude of a triangle* will be defined in the usual way.

§ 19. Parallelograms and Triangles with Equal Bases and Altitudes

Euclid's well-known argument illustrated in the figures below furnishes the following theorem:

THEOREM 44. Two parallelograms with the same bases and with the same altitudes are equicomplementable with each other.

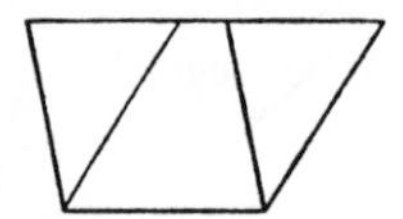

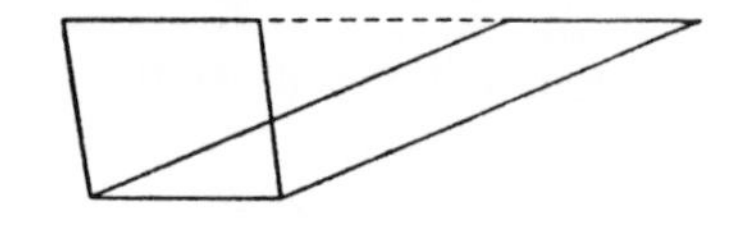

Furthermore the following well-known result also holds:

THEOREM 45. Every triangle ABC is equidecomposable with a parallelogram of an equal base and of half the altitude.

PROOF. Bisecting AC at D and BC at E and then extending DE by an equal amount to itself to F, the resulting triangles DCE and FBE are congruent, and consequently the triangle ABC and the parallelogram $ABFD$ are equidecomposable.

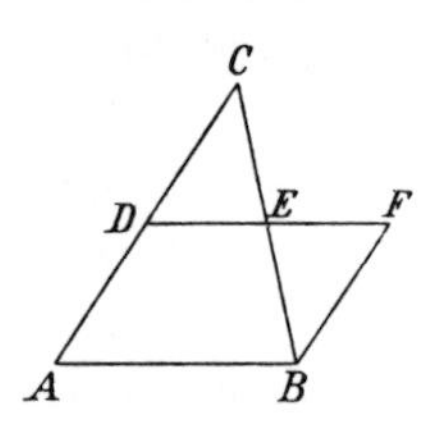

From Theorems 44 and 45 and with the aid of Theorem 43, it follows immediately that:

THEOREM 46. Two triangles with equal bases and altitudes are **equicomplementable.**

As is well known it is easy to show, as the accompanying figure indicates, that two parallelograms, and by Theorems 43 and 45 also two triangles, with equal bases and altitudes are **equidecomposable**. However, it should be noted that *the proof of this without the use of the Archimedean Axiom is impossible.* In fact in every non-Archimedean geometry (for an example of one see Section 12 of Chapter II) it is possible to specify two triangles with equal bases and altitudes and which thus, according to Theorem 46, are **equicomplementable** but nevertheless are **not equidecomposable.**

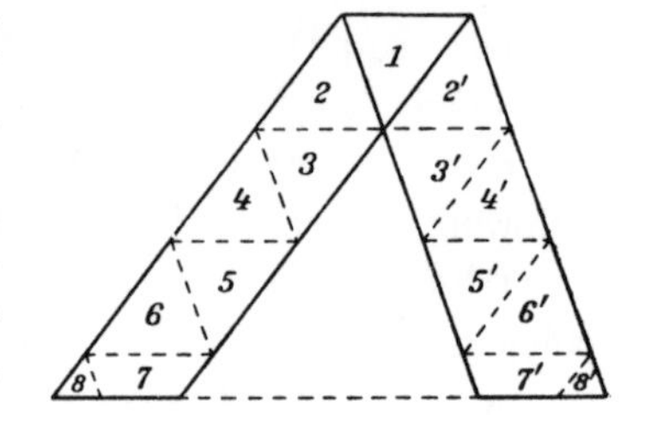

On a ray in a non-Archimedean geometry lay off two segments $AB = e$ and $AD = a$ which for no integer n satisfy the relation

$$n \cdot e \geqq a.$$

Let the perpendiculars AC and DC' of length e be erected at the end points of the segment AD. By Theorem 46 the triangles ABC and ABC' are equicomplementable. From Theorem 23 it follows that the sum of two sides of a triangle are greater than the third side, where the sum of two sides is to be understood in the sense of the segment arithmetic introduced in Chapter III.

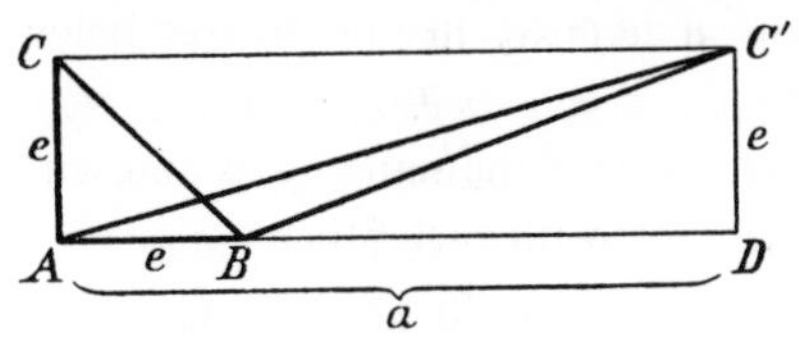

It thus follows that $BC < e + e = 2e$. Furthermore it is possible to prove the following theorem without the use of continuity: A segment lying entirely in the interior of a triangle is smaller than the greatest of its sides. Consequently every segment lying in the interior of the triangle ABC is also less than $2e$.

Assume now that decompositions of the triangles ABC and ABC' into a finite number, say k in each, of congruent triangles in pairs are given. Every side of a partial triangle used in the decomposition of the triangle ABC lies either inside the triangle ABC or on one of its sides, i.e., it is less than $2e$. The perimeter of every partial triangle is therefore less than $6e$. Consequently the sum of all these perimeters is less than $6k \cdot e$. The decomposition of the triangles ABC and ABC' must yield the same sum for the perimeters so the sum of the perimeters used in the decomposition of ABC' must also be less than $6k \cdot e$. In this sum the entire side AC' is certainly a summand, i.e., $AC' < 6k \cdot e$ and hence, by Theorem 23, a fortiori $a < 6k \cdot e$. This contradicts the hypothesis about the segments e and a. The assumption of the possibility to decompose the triangles ABC and ABC' into partial triangles congruent in pairs has thus led to a contradiction.

The important theorems of elementary geometry on equicomplementability of polygons, in particular the Pythagorean Theorem, are easy corollaries of the above-established theorems. The following theorem will be mentioned:

THEOREM 47. For every triangle and hence for every simple polygon it is always possible to construct a right triangle, one whose legs are 1 and which is equicomplementable with the triangle or with the polygon.

The assertion for triangles follows easily by Theorems 43 and 46. The assertion for polygons is obtained as follows: Decompose the given simple polygon into triangles and determine for them equicomplementable right triangles with unit legs in each of them. Since the legs of unit length form the altitudes of these triangles a composition (p. 60) on the basis of Theorems 43 and 46 leads then to the assertion.

In continuing to develop the theory of area an essential difficulty is encountered. In particular, the investigations conducted so far leave in doubt whether perhaps not **all** polygons are equicomplementable. In such a case all theorems established above would become moot and meaningless. With this is connected the question whether in two equicomplementable rectangles with a common side the other sides necessarily are equal.

A closer examination shows that in order to answer the posed question one needs the converse of Theorem 46 which reads as follows:

THEOREM 48. *If two equicomplementable triangles have the same bases then they also have the same altitudes.*

This fundamental theorem is found in the first book of Euclid's Elements as Theorem 39. In the proof Euclid appeals to the general theorem on magnitudes "Καὶ τὸ ὅλον τοῦ μέρους μεῖζόν ἐστιν" (The whole is greater than any of its parts), a method that is equivalent to the introduction of a new geometric axiom of equicomplementability.[1]

However it is possible to establish Theorem 48 and also the theory of area in the manner proposed, i.e., with the aid of the plane axioms alone and without the use of the Archimedean Axiom. In order to see this, one needs the concept of area.

§ 20. The Areas of Triangles and Polygons

DEFINITION. A line AB partitions the points of a plane geometry not lying on it into two regions of points. One of these regions is said to lie to the *right* of the ray AB, emanating from A, or of the "*directed segment* AB" and to lie to the *left* of the ray BA, emanating from B, or of the "*directed segment* BA." The other one is said to lie to the left of the ray AB and to the right of the ray BA. The same region is said to lie to the right of two directed segments AB and AC if B and C lie on the same ray emanating from A (and conversely). Once the right region is

[1] In fact, in Appendix II a geometry will be constructed in which the axioms taken here as a basis, Axioms I-IV, with the exception of Axiom III, 5 for which a more restrictive formulation will be chosen, are all satisfied, and in which nevertheless Theorem 48, and hence also the proposition "The whole is greater than any of its parts" are not valid. Cf. p. 127 ff.

defined for a ray g emanating from O, and if a ray h emanating from O lies in this region then with respect to h the region that contains g is said to lie to the **left** of h. It is apparent that in this way, starting with a definite ray AB, the right and left sides in a plane geometry are uniquely determined with respect to **every** ray or every directed segment.

The points in the interior (p. 9) of a triangle ABC lie either to the left of the sides AB, BC, CA or to the left of CB, BA, AC. In the first case ABC (or BCA, or CAB) is said to be the *positive orientation* and CBA (or BAC, or ACB) is said to be the *negative orientation* of the triangle. In the last case, CBA is said to be the positive orientation and ABC to be the negative orientation of the triangle.

DEFINITION. If the two altitudes $h_a = AD$ and $h_b = BE$ are constructed in a triangle ABC with the sides a, b, c the proportion

$$a : h_b = b : h_a,$$

i.e.,

$$ah_a = bh_b;$$

follows by Theorem 41 from the similarity of the triangles BCE and ACD. Hence in every triangle the product of a base and its corresponding altitude is independent of the side of the triangle that is chosen for a base. Half of the product of the base and the altitude is therefore a segment a that is characteristic of the triangle ABC. Let the orientation of the triangle ABC be positive. The **positive** segment a (according to the definition on p. 57) will be called now the *area of the positively oriented triangle ABC* and denoted by $[ABC]$. The **negative** segment $-a$ will be called the *area of the negatively oriented triangle ABC* and denoted by $[CBA]$.

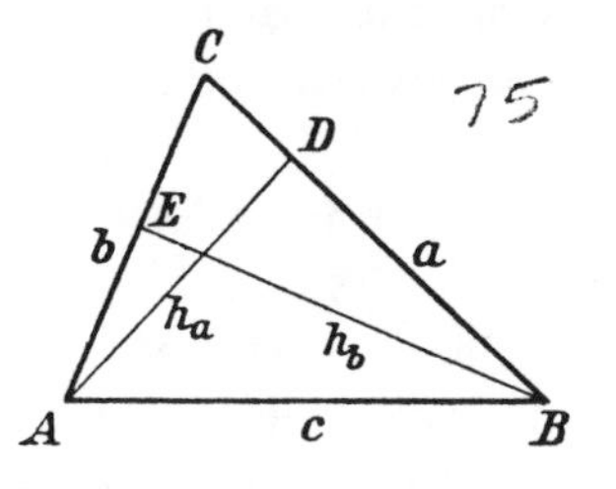

The following simple theorem then holds:

THEOREM 49. If a point O lies outside a triangle ABC then the relation

$$[ABC] = [OAB] + [OBC] + [OCA]$$

holds for the area of the triangle.

PROOF. Assume that the segments AO and BC meet at a point D. Then with the aid of the distributive laws of segment arithmetic one obtains from the definition of area the relations

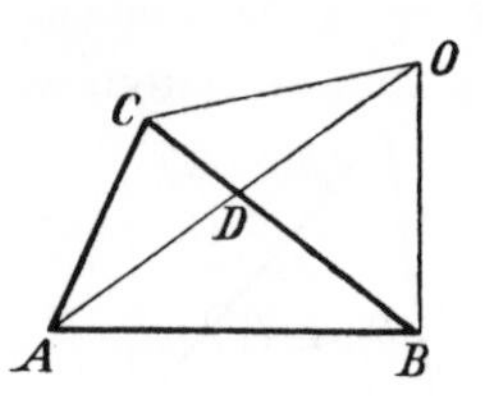

$$[OAB] = [ODB] + [DAB],$$
$$[OBC] = -[OCB] = -[OCD] - [ODB],$$
$$[OCA] = [OCD] + [CAD].$$

The addition of the segments in these equations with the aid of a theorem stated on p. 57 yields

$$[OAB] + [OBC] + [OCA] = [DAB] + [CAD],$$

and hence, again by the distributive law, follows

$$[OAB] + [OBC] + [OCA] = [ABC].$$

The remaining possible assumptions concerning the position of O lead in a corresponding manner to the assertion of Theorem 49.

THEOREM 50. If a triangle ABC is decomposed into a finite number of triangles $\triangle_k$ then the area of the positively oriented triangle ABC is equal to the sum of the areas of all positively oriented triangles $\triangle_k$.

PROOF. Let ABC be the positive orientation of the triangle ABC and let DE be a segment in the interior of the triangle ABC on which the two triangles DEF and DEG border in the decomposition. Let DEF be the positive orientation of the triangle DEF. Then GED is the positive orientation of the triangle DEG. Choosing now a point O outside the triangle ABC the relations

$$[DEF] = [ODE] + [OEF] + [OFD],$$
$$[GED] = [OGE] + [OED] + [ODG],$$
$$= [OGE] - [ODE] + [ODG]$$

hold by Theorem 49. Adding these two segment equations there results the area $[ODE]$ on the right-hand side.

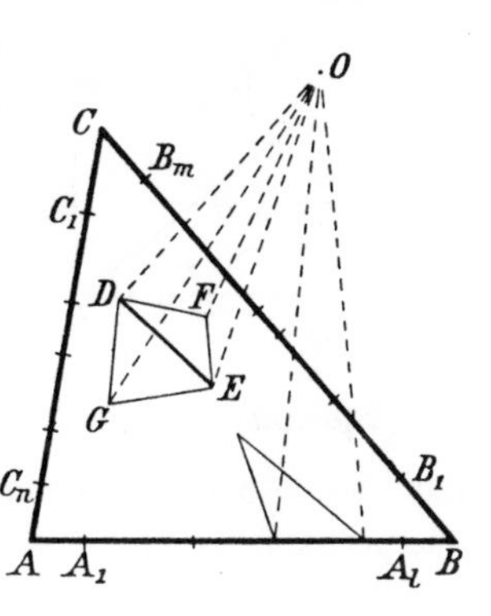

Express the areas of all positively oriented triangles $\triangle_k$ in this manner in accordance with Theorem 49 and sum all segment equations generated in this way. Then for **every** segment DE lying in the interior of the triangle ABC there results the area $[ODE]$ on every right-hand side. Denoting, in correspondence to their ordering, the points used for the decomposition of the triangle ABC lying on its sides by the sequence $A, A_1, \ldots, A_l, B, B_1, \ldots, B_m, C, C_1, \ldots, C_n$ and denoting for short by Σ the sum of the areas of all positively oriented triangles $\triangle_k$ the addition of all segment equations yields, as is easy to see,

$$\begin{aligned}\Sigma &= [OAA_1] + \ldots + [OA_lB] \\ &+ [OBB_1] + \ldots + [OB_mC] \\ &+ [OCC_1] + \ldots + [OC_nA] \\ &= [OAB] + [OBC] + [OCA].\end{aligned}$$

Hence, by Theorem 49 $\qquad \Sigma = [ABC].$

DEFINITION. Let the area $[P]$ of a positively oriented simple polygon be defined as the sum of the areas of all positively oriented triangles into which the polygon splits in some definite decomposition. By an argument similar to the one used in Section 18 for the proof of Theorem 43 it becomes apparent by Theorem 50 that the area $[P]$ is independent of the manner of decomposition into triangles and thus is uniquely determined only by the polygon.

With the aid of Theorem 50 one deduces that **equidecomposable polygons have equal areas**. (Here and in what follows area is always to be interpreted as that for positive orientation.)

Furthermore, if P and Q are equicomplementable polygons then by definition there necessarily exist polygons $P', Q'; \ldots; P'', Q''$ such that the polygon $P + P' + \ldots + P''$ composed of $P, P', \ldots, P''$ is equidecomposable with the polygon $Q + Q' + \ldots + Q''$ composed of

$Q, Q', \ldots, Q''$. From the equations

$$\begin{aligned} [P + P' + \ldots + P''] &= [Q + Q' + \ldots + Q''] \\ [P'] &= [Q'] \\ &\vdots \\ [P''] &= [Q''] \end{aligned}$$

it is easy to conclude that

$$[P] = [Q],$$

i.e., **equicomplementable polygons have the same areas.**

§ 21. Equicomplementability and Area

In Section 20 it was found that equicomplementable polygons have the same areas. From this result the proof of Theorem 48 follows immediately. Denoting the equal bases of the two triangles by g, the corresponding altitudes by h and h' one concludes from the assumed equicomplementability of the two triangles that they also must have the same areas, i.e., it follows that

$$\tfrac{1}{2}gh = \tfrac{1}{2}gh'$$

and hence, after dividing by $\tfrac{1}{2}g$,

$$h = h'.$$

This is the assertion of Theorem 48.

It is also possible now to convert the statement made at the end of Section 20. In fact let P and Q be two polygons with the same areas. In accordance with Theorem 47 construct two right triangles Δ and E of the following sort: Let each have a unit leg and let the triangle Δ be equicomplementable with the polygon P and the triangle E equicomplementable with the polygon Q. It follows then from the theorem proved at the end of Section 20 that Δ and P and also E and Q have the same areas. In view of the equality of the areas of P and Q it follows that Δ and E also have the same areas. Since the two right triangles coincide in their unit legs they necessarily coincide in their other legs, i.e., the triangles are congruent. Hence, by Theorem 43, the two polygons P and Q are equicomplementable.

The two results deduced in the preceding paragraph are combined in the following theorem:

THEOREM 51. *Two equicomplementable polygons have the same areas and two polygons with the same areas are equicomplementable.*

In particular, two equicomplementable rectangles with a common side must also coincide in their other sides. The following also holds:

THEOREM 52. If a rectangle is decomposed by lines into several triangles and one of these triangles is omitted then it is impossible to fill out the rectangle with the remaining triangles.

This theorem was taken as an axiom by de Zolt[1] and O. Stolz[2] and was proved by F. Schur[3] and W. Killing[4] with the aid of the Archimedean Axiom. In the foregoing discussion it has been shown **that it is completely independent of the Archimedean Axiom.**

In essence, the segment arithmetic introduced in Section 15 of Chapter III was used to prove Theorems 48, 50, 51 and since these rest essentially on Pascal's Theorem (Theorem 40) or rather on a special case of it (p. 49) Pascal's Theorem emerges as the most important building block in the theory of area.

It is easy to see that, conversely, Pascal's Theorem can be deduced from Theorems 46 and 48. In fact, from the parallelism of the lines CB' and $C'B$ the equicomplementability of the triangles OBB' and OCC' follows from Theorem 48. Also the equicomplementability of the triangles OAA' and OCC' follows from the parallelism of the lines CA' and AC'. Since according to this the triangles OAA' and OBB' are also equicomplementable Theorem 48 shows that BA' and AB' must also be parallel.

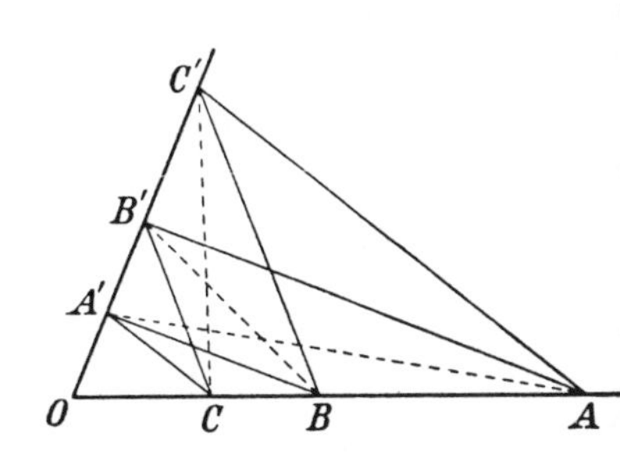

Furthermore, it is easy to see that the area of a polygon that lies completely in the interior of another polygon is less than that of the

[1] *Principii della equaglianza di poligoni preceduti da alcuni critici sulla teoria della equivalenza geometrica* (Milano, Briola, 1881). Cf. also *Principii della equaglianza di poliedri e di poligoni sferici* (Milano, Briola, 1883).

[2] *Monatshefte für Math. und Phys.*, Jahrgang 5 (1894).

[3] *Sitzungsberichte der Dorpater Naturf. Ges.*, 1892.

[4] *Grundlagen der Geometrie*, Vol. 2, Part 5, Section 5 (1898).

latter, and thus by Theorem 51 cannot be equicomplementable with it. This result contains Theorem 52 as a special case.

The essential theorems of the theory of plane area have thus been established.

Gauss had already drawn the attention of mathematicians to the same problem in space. I have expressed a conjecture concerning the impossibility of an analogous development of the theory of volume in space, and posed the specific problem[1] of finding two tetrahedra with equal base areas and equal altitudes which could in no way be decomposed into congruent tetrahedra and which by the adjunction of congruent tetrahedra could not be expanded to polyhedra which in turn could be decomposed into congruent tetrahedra.

M. Dehn[2] has in fact proved this conjecture. At the same time, he proved more rigorously the impossibility of developing the theory of volume in the same way as the theory of plane areas was developed in the foregoing paragraphs.

Hereafter in order to treat analogous problems in space, other means, such as the Cavalieri principle, would have to be resorted to.[3]

W. Süss[4] has developed in this sense the theory of volume in space. He calls two tetrahedra of equal altitudes and of equicomplementable base areas equal in the sense of Cavalieri. He also calls two polyhedra equidecomposable in the sense of Cavalieri if they can be decomposed into a finite number of tetrahedra that are equal in the sense of Cavalieri. Finally two polyhedra which can be represented by differences of polyhedra that are equidecomposable in the sense of Cavalieri he calls equicomplementable in the sense of Cavalieri. It is possible to show without the use of the continuity axiom that equality

[1] Cf. my lecture "Mathematische Probleme," No. 3.

[2] "Über raumgleiche Polyeder," *Göttinger Nachrichten*, 1900, as well as "Über den Rauminhalt," *Math. Ann.*, Vol. 55 (1902). Cf. further Kagan, *Math. Ann.*, Vol. 57.

[3] Only the first part of Theorem 51, as well as Theorem 48 and Theorem 52, hold analogously in space. Cf. Schatunowsky, "Über den Rauminhalt der Polyeder," *Math. Ann.*, Vol. 57. In the exposition "Über den Inhalt sphärischer Dreiecke," *Math. Ann.*, Vol. 60, M. Dehn has shown that with the aid of the congruence theorems it is possible to develop the theory of plane area without the axiom of parallels. See further Finzel, "Die Lehre vom Flächeninhalt in der allgemeinen Geometrie," *Math. Ann.*, Vol. 72.

[4] "Begründung der Lehre vom Polyederinhalt," *Math. Ann.*, Vol. 82.

of area and equicomplementability in the sense of Cavalieri are equivalent concepts, while the equidecomposability in the sense of Cavalieri in polyhedra of equal volume can only be proved with the aid of the Archimedean Axiom.

As a recent result obtained by J.P. Sydler[1] the following is mentioned: The plane theorem deduced from Theorem 51 and from the discussion on p. 62 (subsequent to Theorem 46) that two equicomplementable polygons are also equidecomposable, can be extended to polyhedra in space by assuming the Archimedean Axiom. From this result one deduces in turn that the equivalence classes of polyhedra with respect to equidecomposability has the power of the continuum.

[1] "Sur la décomposition des polyèdres," *Comm. Helv.*, Vol. 16 (1943/44), 266-73.

CHAPTER V

DESARGUES' THEOREM

§ 22. Desargues' Theorem and Its Proof in the Plane with the Aid of the Congruence Axioms

Among the axioms formulated in Chapter I, those of groups II-V are in part line axioms and in part plane axioms. Axioms 4-8 of group I are the only space axioms. In order to grasp clearly the meaning of the space axioms let some **plane** geometry be assumed and the general conditions be studied under which this geometry can be embedded in a space geometry in which the axioms formulated for the plane geometry, and also the space incidence Axioms I, 4-8, are all satisfied.

In this and in the following chapters the congruence axioms are generally not resorted to. Consequently the axiom of parallels IV (p. 25) must be laid down in sharper form.

IV* **(Axiom of parallels in sharper form).** *Let a be any line and A a point off a. Then there exists in the plane determined by a and A* **one and only one** *line that passes through A and does not intersect a.*

It is well known that the so-called Theorem of Desargues can be proved on the basis of the axioms of groups I-II, IV*. Desargues' Theorem is a plane intersection theorem. The line on which the points of intersection of the corresponding sides of the two triangles lie is particularly distinguished, and is customarily referred to as "the line at infinity." The theorem and its converse resulting in this case will be called Desargues' Theorem. This theorem reads as follows:

THEOREM 53 **(Desargues' Theorem).** If two triangles are situated in a plane so that pairs of corresponding sides are parallel then the lines joining the corresponding vertices pass through one and the same point or are parallel.

Conversely, if two triangles lie in a plane so that the lines joining corresponding vertices pass through one point or are parallel and further, if two pairs of corresponding sides of the triangles are parallel then the third sides of the two triangles are also parallel.

As already pointed out, Theorem 53 is a consequence of Axioms I,

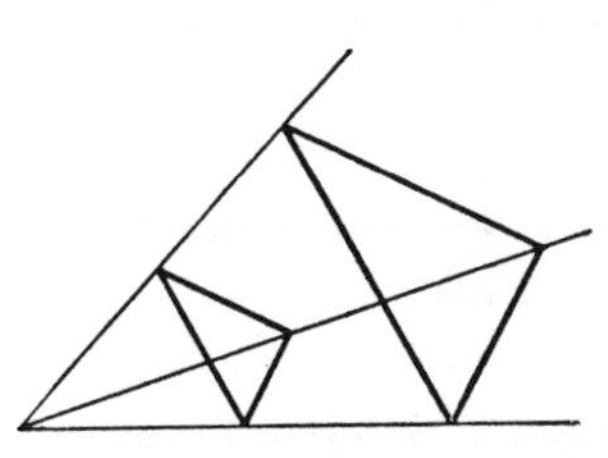

II, IV*. According to this result, the validity of Desargues' Theorem in a plane geometry is at any rate a **necessary** condition for this geometry to be embeddable in a space geometry in which the axioms of groups I, II, IV* are all satisfied.

As in Chapter III and IV let a **plane** geometry be assumed in which Axioms I, 1-3 and II-IV hold, and suppose that a segment arithmetic is introduced in accordance with Section 15. Then, as it has been shown in Section 17, it is possible to assign to every point of the plane a pair of segments (x, y) and to every line a ratio of three segments $(u : v : w)$ in which u, v are not both zero so that the linear equation

$$ux + vy + w = 0$$

represents the condition for the common position of the point and of the line. According to Section 17 the set of all lines in this geometry forms a number field for which properties 1-16 enumerated in Section 13 hold. Hence, as was done in Section 9 and in Section 12 by means of the number sets Ω and $\Omega(t)$, respectively, it is possible to construct a space geometry. To do this it is stipulated that a set of three segments (x, y, z) represent a point, the ratios of four segments $(u : v : w : r)$ in which u, v, w do not vanish simultaneously represent a plane while the lines be defined as the intersections of pairs of planes. In this manner the linear equation

$$ux + vy + wz + r = 0$$

expresses the fact that the point (x, y, z) lies in the plane $(u : v : w : r)$. Finally, concerning the ordering of points on a line, or of points in a plane with respect to a line in it, or the ordering of points in space with respect to a plane, these can be done by the inequalities among the segments in an analogous manner to that carried out for the plane in Section 9.

Since the original plane geometry can be recovered by setting $z = 0$ it becomes apparent that this plane geometry can be regarded as a part of a space geometry. For this, according to the above discussions, the validity of Desargues' Theorem is a necessary condition, and it follows hence that Desargues' Theorem must also hold in the assumed plane geometry. It is then a consequence of Axioms I, 1-3, II-IV.

Note that the result just found can also be obtained without difficulty directly from Theorem 42 in the theory of proportion, or from Theorem 61.

§ 23. The Impossibility of Proving Desargues' Theorem in the Plane without the Aid of the Axioms of Congruence

In studying the question whether it is possible to prove Desargues' Theorem in plane geometry without the axioms of congruence the following result is reached:

THEOREM 54. *There exists a plane geometry in which Axioms* I, 1-3, II, III, 1-4, IV*, *i.e., all line and plane axioms with the exception of the congruence Axiom* III, 5, *hold but in which Desargues' Theorem* (Theorem 53) **does not** *hold. Therefore Desargues' Theorem* **cannot be deduced** *from the above-mentioned axioms alone. For its proof either the space axioms, or Axiom* III, 5 *about the congruence of triangles are necessary.*

PROOF.[1] In the ordinary plane Cartesian geometry, whose existence has already been shown in Section 9 of Chapter II, change the definitions of lines and angles in the following way: Choose any line in the Cartesian geometry as an axis and introduce positive and negative directions on this axis as well as positive and negative half planes with respect to this axis.

As lines in this new geometry take this axis and every line in the Cartesian geometry that is parallel to it, every line in the Cartesian geometry whose part in the positive half plane forms a right or acute angle with the positive direction of the axis and finally, every pair of rays h, k with the property that the common vertex of h and of k lie on the axis. Let the ray h lying in the positive half plane form with the positive direction of the axis an acute angle α and let the extension k' of the ray k that lies in the negative half plane form with the positive direction of the axis an angle β so that in the Cartesian geometry the relation

$$\frac{\tan \beta}{\tan \alpha} = 2$$

holds.

[1] Instead of the first "non-Desarguesian" geometry introduced at this point in the previous editions of this book, a somewhat simpler non-Desarguesian geometry due to Moulton will be illustrated subsequently. Cf. R. F. Moulton, "A Simple non-Desarguesian Plane Geometry," *Trans. Am. Math. Soc.*, 1902.

The ordering of points and the lengths of segments are to be defined in the usual manner also on the lines which in the Cartesian geometry are represented by pairs of rays. It can easily be seen that in the geometry defined in this manner Axioms I, 1-3, II, III, 1-3, IV* are valid. For example, it can be seen immediately that the lines passing through one point cover the plane in a simple manner. Besides, Axioms V are also valid.

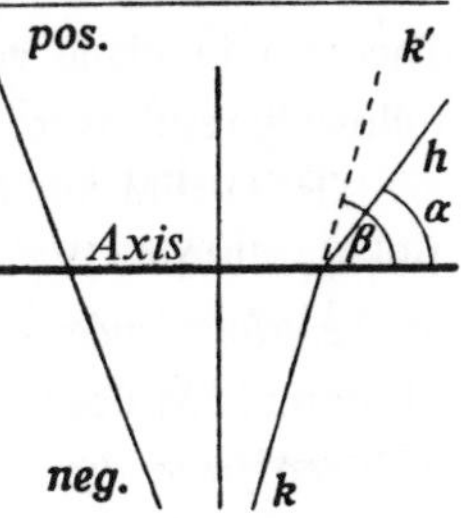

All angles, which do not have at least **one** side that emanates from the axis into the positive half plane and forms an acute angle with the positive direction of the axis, are measured as in Cartesian geometry in the usual way. However, if at least one side of an angle ω is a ray h with the just-mentioned properties, then the value of the angle ω in the new geometry is defined as the value of the angle ω' in the Cartesian geometry which instead of h has the corresponding ray k' (see the preceding page) as a side. The figure on the left illustrates this method of definition for two pairs of supplementary angles.

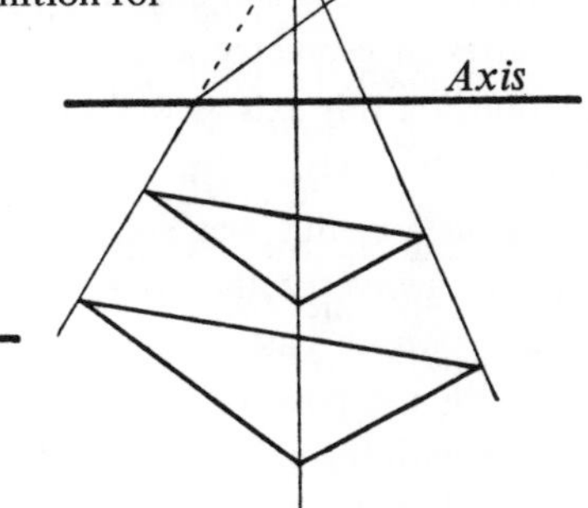

On the basis of this definition Axiom III, 4 is also valid. In particular, for every angle $\sphericalangle(l, m)$

$$\sphericalangle(l, m) \equiv \sphericalangle(m, l).$$

However, as the figure on the right shows at once, and as can be verified by calculation, **Desargues' Theorem does not hold in this new plane geometry**. It is just as easy to draw a figure that would show that Pascal's Theorem does not hold either.

The plane "non-Desarguesian" geometry given here serves at the same time as an example of a plane geometry in which Axioms I, 1-3, II, III, 1-4, IV* are valid and which nevertheless can **not** be embedded in a space geometry.[1]

§ 24. Introduction of a Segment Arithmetic Based on Desargues' Theorem with the Aid of the Congruence Axioms [2]

In order to appreciate fully the significance of Desargues' Theorem (Theorem 53) a plane geometry will be constructed in which Axioms I, 1-3, II, IV*[3], i.e., all line and plane axioms except the congruence and continuity axioms, are valid. In this geometry a new segment arithmetic, **independent of the congruence axioms**, will be introduced in the following way:

Take two fixed lines in the plane which intersect at the point O and consider only such segments whose initial points are O and whose end points lie anywhere on these two fixed lines. Denote the point O itself as the segment 0. Symbolically,

$$OO = 0 \quad \text{or} \quad 0 = OO.$$

Let E and E' be fixed points each on one of two fixed lines through O. Denote the two segments OE and OE' as the segment 1. Symbolically,

$$OE = OE' = 1$$

or

$$1 = OE = OE'.$$

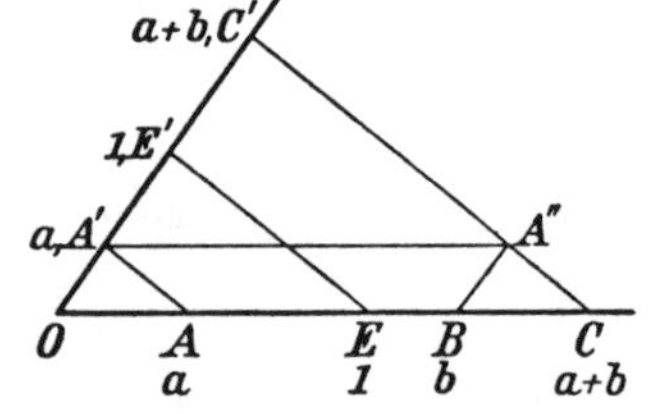

Call the line EE', the **unit line** for short. Furthermore, if A and A' are points on the lines OE and OE', respectively, and if the connecting line AA' is parallel to EE' consider the segments OA and OA' to be equal.

[1] H. Mohrmann gives further interesting examples of non-Desarguesian line systems in *Festschrift David Hilbert* (Berlin, 1922), p. 181.

[2] A derivation of a segment arithmetic in connection with the development of the ideas of topology is given by G. Hessenberg in his article "Über einen geometrischen Kalkül," *Acta Math.*, Vol. 29, 1904. Some parts of the derivation follow more easily if vector addition in the plane is developed on the basis of Desargues' Theorem. Cf. Hölder, *Streckenrechnung und projektive Geometrie* (Leipzig and Berlin, 1911).

[3] A new segment arithmetic can be introduced by a projective form of Desargues' Theorem even without the axiom of parallels IV*. Concerning the possibility of dropping the order axioms, cf. Suppl. IV.

Symbolically,

$$OA = OA' \quad \text{or} \quad OA' = OA.$$

In order to define the sum of the segments $a = OA$ and $b = OB$ lying on OE construct AA' parallel to the unit line EE' and draw a parallel to OE through A', and through B a parallel to OE'. These two parallels intersect at a point A''. Finally draw a parallel to the unit line EE' through A''. It meets the fixed lines OE and OE' at the points C and C', respectively. Then call $c = OC = OC'$ the *sum* of the segment $a = OA$ and the segment $b = OB$. Symbolically,

$$c = a + b \quad \text{or} \quad a + b = c.$$

Let it be supposed that by assuming the validity of Desargues' Theorem (Theorem 53) the sum of two segments can generally be obtained. The point C which determines the sum $a + b$ on the line on which A and B lie is then independent of the adopted unit line EE', i.e., the point C can also be obtained by the following construction:

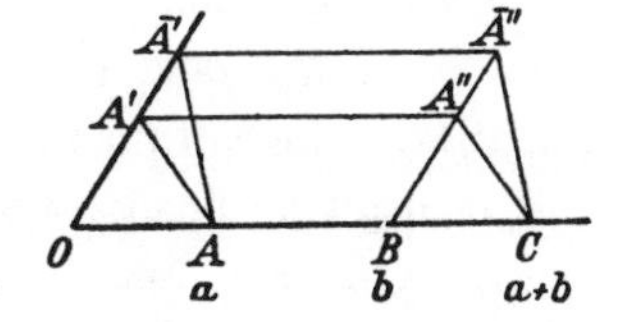

Choose any point $\bar{A}'$ on the line OA' and draw a parallel to $O\bar{A}'$ through B and the parallel to OB through $\bar{A}'$. These two parallels meet at a point $\bar{A}''$. The parallel to $A\bar{A}'$ drawn now through $\bar{A}''$ meets the line OA at the point C which determines the sum $a + b$.

For the proof it will be assumed that the points A' and A'' as well as the points $\bar{A}'$ and $\bar{A}''$ are obtained in the outlined manner and that the point C is determined on OA in such a way that CA'' is parallel to AA'. Then it is necessary to show that $C\bar{A}''$ is also parallel to $A\bar{A}'$. The triangles $AA'\bar{A}'$ and $CA''\bar{A}''$ are situated in such a way that the connecting lines of corresponding vertices are parallel and consequently two pairs of corresponding sides, namely $A'\bar{A}'$ and $A''\bar{A}''$ as well as AA' and CA'', are parallel. In fact by the second assertion of Desargues' Theorem the third sides $A\bar{A}'$ and $C\bar{A}''$ are also parallel.

In order to define the product of a segment $a = OA$ and a segment $b = OB$ the same construction outlined in Section 15 will be utilized,

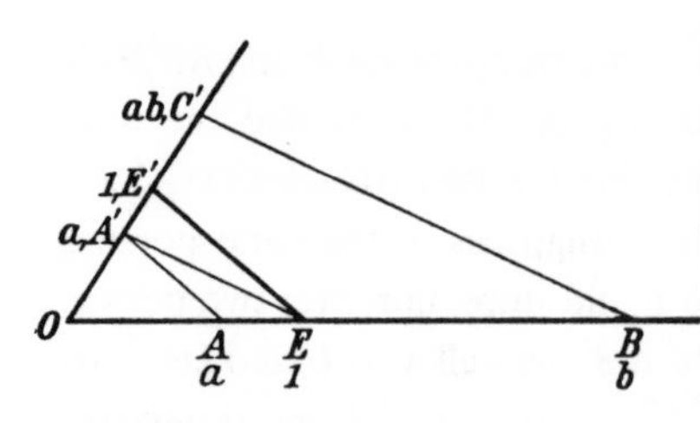

except that now the sides of the right angle will be replaced by the two fixed lines OE and OE'. Consequently the construction is as follows: Determine the point A' on OE' so that AA' becomes parallel to the unit line EE'. Join E with A' and through B draw a parallel to EA'. This parallel meets the fixed line OE' at a point C'. Call $c = OC'$ the *product* of the segment $a = OA$ and the segment $b = OB$. Symbolically,

$$c = ab \quad \text{or} \quad ab = c.$$

§ 25. The Commutative and the Associative Laws of Addition in the New Segment Arithmetic

As is easy to see, all theorems of composition formulated in Section 13 are valid for this new segment arithmetic. We will now investigate which of the rules of arithmetic formulated there are valid in this segment arithmetic when one starts out with a plane geometry in which Axioms I, 1-3, II, IV* are satisfied, and **Desargues' Theorem holds as well.**

It will be shown first that the **commutative** law of addition of segments

$$a + b = b + a$$

defined in Section 24, holds. Let

$$a = OA = OA',$$
$$b = OB = OB',$$

whereby, according to the definition employed, AA' and BB' are parallel to the unit line. Construct the point A'' and B'' by drawing $A'A''$ as well as $B'B''$ parallel to OA, and also AB'' and BA'' parallel to OA'. As can be seen immediately the assertion states that the connecting line $A''B''$ runs parallel to AA'. The truth of this assertion becomes apparent by Desargues' Theorem (Theorem 53) as follows: Denote by F the point of intersection of AB'' with $A'A''$ and the point of intersection of BA'' with $B'B''$

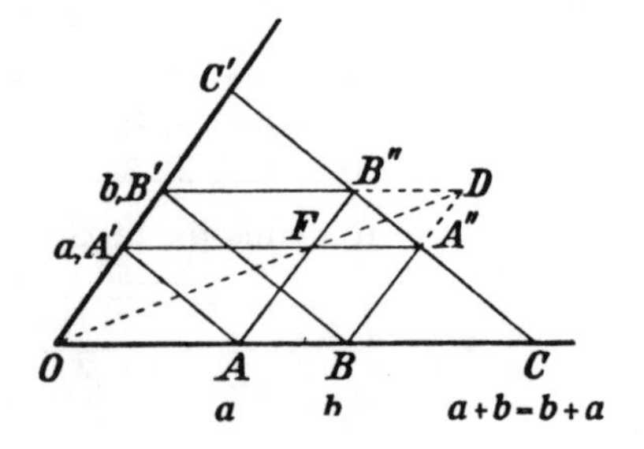

by D. Then the corresponding sides of the triangles $AA'F$ and $BB'D$ are parallel. One concludes then by Desargues' Theorem that the three points O, F, D are collinear. In this case the two triangles OAA' and $DB''A''$ are so situated that the connecting lines of the corresponding vertices pass through the same point F and since moreover two pairs of corresponding sides, namely, OA and DB'' as well as OA' and DA'', are parallel, the third sides AA' and $B''A''$ are also parallel by the second assertion of Desargues' Theorem (Theorem 53).

It follows at the same time from this proof that it is immaterial on which one of the two fixed lines the construction of the sum of two segments is started.

Furthermore, the **associative** law of addition

$$a + (b + c) = (a + b) + c$$

holds.

Let the segments

$$a = OA, \quad b = OB, \quad c = OC$$

be given on the line OE. On the basis of the general rules of addition outlined in the preceding section the sums

$$a + b = OG, \quad b + c = OB', \quad (a + b) + c = OG'$$

can be constructed as follows: Choose arbitrarily a point D on the line OE' and join it with A and B. The parallel to OA through D will meet the two parallels to OD through B and through C at the points F and D', respectively. The parallels to AD and BD drawn through F and D' meet now the line OA at the above-mentioned points G and B', respectively, and the parallel to GD drawn through D' meets the line OA at the point G' also mentioned above. The sum $a + (b + c)$ will finally be obtained by drawing the parallel to OD through B' which is met by the lines DD' at a point F', and by drawing a parallel to AD through F'. It thus remains to be shown that $G'F'$ is parallel to AD.

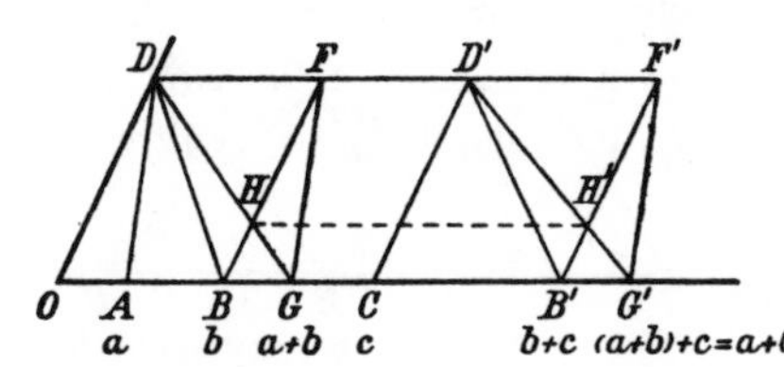

Denoting by H the points of intersection of the line BF with GD and by H' the point of intersection of the line $B'F'$ with $G'D'$ it follows that corresponding sides of the triangles BDH and $B'D'H'$ are parallel. Furthermore, since the two lines BB' and DD'

are parallel, by Desargues' Theorem the line HH' is also parallel to these two lines. It is thus possible to apply the second assertion of Desargues' Theorem to the triangles GFH and $G'F'H'$ and hence realize that $G'F'$ is parallel to GF and, in fact, is also parallel to AD.

§ 26. The Associative Law of Multiplication and the Two Distributive Laws in the New Segment Arithmetic

With the assumptions made the **associative** law of multiplication of segments

$$a(bc) = (ab)c$$

also holds.

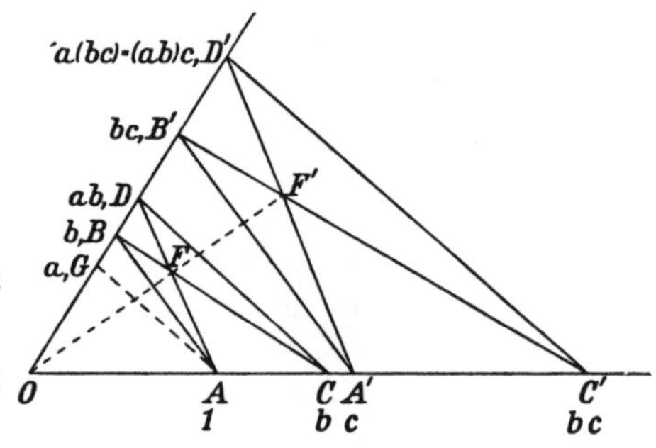

Let the segments

$1 = OA, \ b = OC,$

$c = OA'$

be given on one of the two fixed lines through O and the segments

$a = OG$ and $b = OB$

on the other. In order to construct the segments

$$bc = OB' \quad \text{and} \quad bc = OC',$$
$$ab = OD$$
$$(ab)\,c = OD'$$

according to Section 24 draw $A'B'$ parallel to AB, $B'C'$ parallel to BC, CD parallel to AG and $A'D'$ parallel to AD. It can be seen at once that the assertion is equivalent to saying that CD too must be parallel to $C'D'$. Denoting by F the point of intersection of the line AD with BC and by F' the point of intersection of the line $A'D'$ with $B'C'$ the corresponding sides of the triangles ABF and $A'B'F'$ are parallel. Hence by Desargues' Theorem the three points O, F, F' are collinear. In view of this situation the second assertion of Desargues' Theorem can be applied to the triangles CDF and $C'D'F'$ and then it can be seen that CD is in fact parallel to $C'D'$.

Finally the two distributive laws

$$a(b + c) = ab + ac$$

and

$$(b + c)a = ba + ca$$

will be demonstrated with Desargues' Theorem.

For the proof of the **first distributive law**

$$a(b + c) = ab + ac$$

assume that the segments

$$1 = OE, \quad b = OB, \quad c = OC$$

are given on the first of the two fixed lines and that the segment

$$a = OA$$

is given on the second.

The lines parallel to EA drawn through B and C meet the line OA at the points D and F, respectively. Then by the rule of multiplication of Section 24

$$OD = ab, \quad OF = ac.$$

According to the generalized rule of addition of Section 24 one obtains the sum

$$OH = b + c,$$

by drawing the parallels to OD through C and to OC through D and also by drawing through the point of intersection G of these two lines the parallel to BD which meets OC at the point H mentioned above, and OD at a point K. Since $OH = b + c$, then by the rule of multiplication,

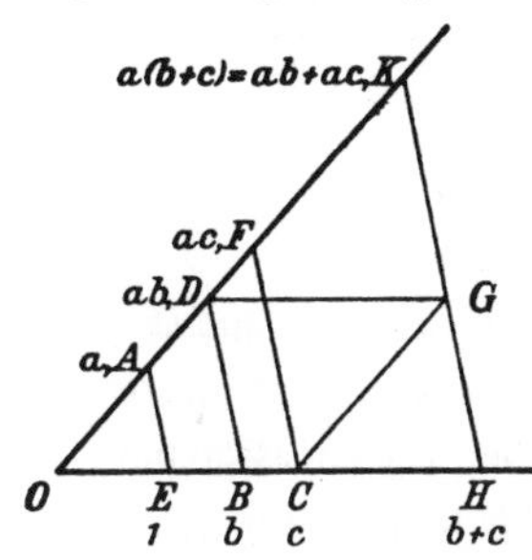

$$OK = a(b + c).$$

According to the generalized rule of addition and the interchangeability of the fixed lines OE, OE' in the construction of a sum proved on p. 78 the sum $ac + ab$ can finally be constructed in the following way: Through any point of OE, say through C, draw the parallel CG to OD, the parallel DG to OC through D and the parallel GK to CF through G.

Then

$$OK = ac + ab$$

and hence follows, with the aid of the commutative law of addition, the first distributive law.

In order to prove the **second distributive law** assume that the segments

$$1 = OE, \quad a = OA$$

are given on the first of the two fixed lines and that the segments

$$b = OB, \qquad c = OC$$

are given on the second. The segments

$$OB' = ba, \qquad OC' = ca$$

are determined by the parallels AB' to EB and AC' to EC respectively. Construct the segments

$$OF = b + c, \quad OF' = ba + ca$$

again on the fixed line OB by the generalized rule of addition as follows: Draw the parallels to OE through C and to OC through E. Through the point D at which they meet draw the parallel to EB which is met by OA at the point F mentioned above. In the same way draw the parallels to OC' through A and to OA through C'. Through the point D' at which they meet draw the parallel to AB' which is met by OA at the point F' mentioned above.

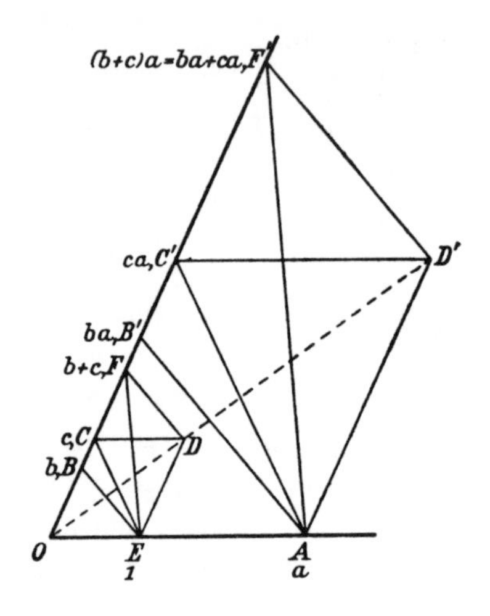

According to the rule of multiplication the second distributive law is proved as soon as it is shown that AF' is parallel to EF.

The corresponding sides of the triangles ECD and $AC'D'$ are parallel. By Desargues' Theorem the three points O, D, D' are then collinear. Therefore it is possible to apply the second assertion of Desargues' Theorem to the triangles EDF and $AD'F'$ and to realize that AF' is in fact parallel to EF.

§ 27. The Equation of a Line Based on the New Segment Arithmetic

In Sections 24-26, a segment arithmetic was introduced by means of the axioms formulated in Section 24 and under the assumption of the validity of Desargues' Theorem in the plane. In this arithmetic, besides the composition theorems formulated in Section 13, the commutative law of addition, the associative laws of addition and multiplication as well as both distributive laws hold. That the commutative law of multiplication does not necessarily persist will become evident in Section 33. In this section it will be shown in what way an analytic representation, based on this segment arithmetic, is possible for points and lines in the plane.

DEFINITION. Let the two fixed lines through O in the plane be termed the X-axis and the Y-axis. Consider any point P in the plane to be determined by the segments x, y which are obtained on the X-axis and the Y-axis, respectively, by drawing parallels to these axes through P. These segments x, y are called the *coordinates* of the point P.

By the new segment arithmetic and Desargues' Theorem one arrives at the following result:

THEOREM 55. *The coordinates* x, y *of the points on any line satisfy a segment equation of the form*

$$ax + by + c = 0;$$

in this equation the segments a, b *must lie* **to the left** *of the coordinates* x, y. *The segments* a, b *are never both zero and* c *is any segment.*

Conversely, every segment equation of the described form represents a line in the adopted plane geometry.

PROOF. The abscissa x of any point P on the Y-axis or on a line parallel to it is independent of the choice of the point P on the desired line, i.e., such a line can be represented in the form

$$x = \bar{c}.$$

For $\bar{c}$ there exists a segment c such that

$$\bar{c} + c = 0$$

and hence

$$x + c = 0.$$

This equation is of the desired form.

Now let l be a line that intersects the Y-axis at a point S. Through any point P on this line draw a parallel to the Y-axis which the X-axis meets at point Q. The segment $OQ = x$ is the abscissa of P. The parallel to l through Q delineates a segment OR on the Y-axis and by the definition of mulitipication it follows that

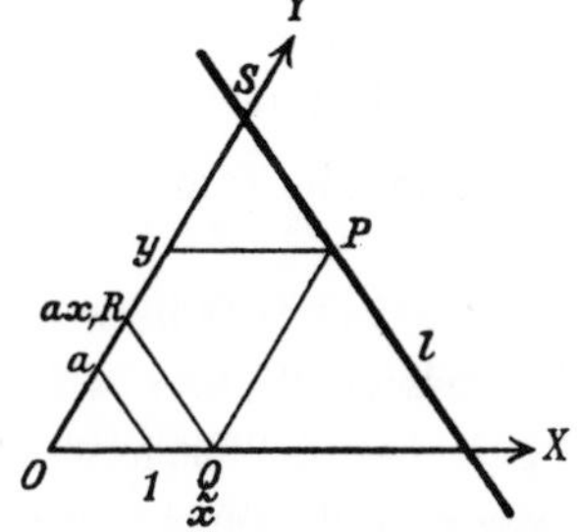

$$OR = ax,$$

where a is a segment which depends only

on the position of l but not on the choice of P on l. Let the ordinate of P be y. According to the extended definition of a sum outlined on pp. 76-78 and in view of the possibility of constructing a sum by starting with the Y-axis, as already shown on p. 78, the segment OS yields now the sum $ax + y$. $OS = \bar{c}$ is a segment determined only by the position of l. From the equation

$$ax + y = \bar{c}$$

follows

$$ax + y + c = 0,$$

where c again is the segment determined by the equation $\bar{c} + c = 0$. The last equation is of the desired form.

It is easy to see that the coordinates of a point not on l do not satisfy this equation.

It is just as easy to prove the validity of the second assertion of Theorem 55. Indeed, given a segment equation

$$a'x + b'y + c' = 0$$

in which not both a' and b' vanish then, if $b' = 0$, multiply the equation on the left by the segment a determined by the relation $aa' = 1$, and if $b' \neq 0$ multiply it by the segment b determined by the relation $bb' = 1$. Then, by the rules of the arithmetic, one obtains an equation of a line as derived above and it is easy to construct in the adopted geometry a line that satisfies this equation.

Let it also be explicitly remarked that under the assumptions made a segment equation of the form

$$xa + yb + c = 0,$$

in which the segments a, b are **to the right** of the coordinates x, y generally does *not* represent a line.

In Section 30 an important application of Theorem 55 will be made.

§ 28. The Totality of Segments Regarded as Complex Numbers

It has already been mentioned that in the new segment arithmetic laid down in Section 24 Theorems 1-6 of Section 13 are valid.

It has also been shown in Sections 25 and 26 with the aid of Desargues' Theorem that in this segment arithmetic, Rules of Operation

7-13 of Section 13 are valid. Thus all theorems of composition and rules of operation hold with exception of the commutativity of multiplication.

Finally, in order to make it possible to order the segments, the following convention will be adopted: Let A, B denote two distinct points on the line OE. Arrange the four points O, E, A, B according to Theorem 5 in a sequence in which E is behind O. If in this sequence B is also behind A, then let it be said that the segment $a = OA$ is *smaller* than the segment $b = OB$.
Symbolically,

$$a < b.$$

However, if in this sequence A is behind B let it be said that the segment $a = OA$ is *greater* than the segment $b = OB$. Symbolically,

$$a > b.$$

It is easy to see that in this arithmetic, Rules of Operation 13-16 are now valid by Axiom II. Thus the totality of all different segments forms a set of complex numbers which has the properties 1-11, 13-16 of Section 13, i.e., **all rules except the commutative law of multiplication and the continuity axioms, are valid.** In what follows such a set of numbers will be called briefly a *Desarguesian set of numbers.*

§ 29. Construction of a Space Geometry with the Aid of a Desarguesian Set of Numbers

Let some Desarguesian set of numbers D be given now. **With it, it will be possible to construct a space geometry in which Axioms I, II, IV* are satisfied.**

In order to see this, consider any set of three numbers, (x, y, z) of a Desarguesian set of numbers D as a point, and any set of four numbers $(u : v : w : r)$ in D, in which the first three numbers are not simultaneously 0, as a plane. However, let the sets $(u : v : w : r)$ and $(au : av : aw : ar)$, where a is some number in D different from 0, represent the same plane. Let the existence of the equation

$$ux + vy + wz + r = 0$$

express the fact that the point (x, y, z) lies in the plane $(u : v : w : r)$.

Finally, a line is defined with the aid of a set of two planes $(u' : v' : w' : r')$ and $(u'' : v'' : w'' : r'')$ when it is possible to find in D a number a distinct from zero such that

$$au' = u'', \quad av' = v'', \quad aw' = w''$$

simultaneously. A point (x, y, z) will be said to lie on the line

$$[(u' : v' : w' : r'), \ (u'' : v'' : w'' : r'')]$$

if it is common to both planes $(u' : v' : w' : r')$ and $(u'' : v'' : w'' : r'')$. Two lines that contain the same points are not to be considered as being distinct.

By applying Rules of Operations 1-11 of Section 13, which by hypothesis are valid in D, it is easy to arrive at the conclusion that Axioms I and IV* hold in the space geometry constructed above.

In order to satisfy also Axioms of Order II, the following conventions will be adopted: Let

$$(x_1, y_1, z_1), \ (x_2, y_2, z_2), \ (x_3, y_3, z_3)$$

be any three points on the line

$$[(u' : v' : w' : r'), \quad (u'' : v'' : w'' : r'')].$$

Then let the point (x_2, y_2, z_2) be said to lie between the other two if at least one of the six pairs of inequalities

$$x_1 < x_2 < x_3, \quad x_1 > x_2 > x_3, \tag{1}$$

$$y_1 < y_2 < y_3, \quad y_1 > y_2 > y_3, \tag{2}$$

$$z_1 < z_2 < z_3, \quad z_1 > z_2 > z_3 \tag{3}$$

is satisfied. Now if one of the two double inequalities (1) holds then it is easy to infer that either $y_1 = y_2 = y_3$ or that one of the double inequalities (2) must hold and that $z_1 = z_2 = z_3$ as well or that one of the double inequalities (3) must hold. In fact by multiplying the equations

$$u'x_i + v'y_i + w'z_i + r' = 0,$$
$$u''x_i + v''y_i + w''z_i + r'' = 0,$$
$$(i = 1, 2, 3)$$

on the left by suitable numbers from D that are distinct from zero and then adding the resulting equations one obtains a set of equations of the form

$$u'''x_i + v'''y_i + r''' = 0 \quad (i = 1, 2, 3). \tag{4}$$

Here the coefficient v''' is certainly distinct from 0 as otherwise the equality of the three numbers x_1, x_2, x_3 would follow. If $u''' = 0$ then one obtains

$$y_1 = y_2 = y_3.$$

However, if $u''' \neq 0$, then one infers from

$$x_1 \lesseqgtr x_2 \lesseqgtr x_3$$

the double inequality

$$u''' x_1 \lesseqgtr u''' x_2 \lesseqgtr u''' x_3$$

and hence, in view of (4)

$$v''' y_1 + r''' \lesseqgtr v''' y_2 + r''' \lesseqgtr v''' y_3 + r'''$$

whence

$$v''' y_1 \lesseqgtr v''' y_2 \lesseqgtr v''' y_3,$$

and since v''' is not 0

$$y_1 \lesseqgtr y_2 \lesseqgtr y_3.$$

In each of these double inequalities either the upper or the lower sign should be regarded as holding throughout.

The preceding discussions show that in this geometry, line axioms of order II, 1-3 hold. It remains to be shown that in it the plane axiom II, 4 is valid.

To do this let a plane $(u : v : w : r)$ and a line $[(u : v : w : r), (u' : v' : w' : r')]$ in it be given. Suppose that all points (x, y, z) lying in the plane $(u : v : w : r)$ lie on one or the other side of the line according to whether the expression $u'x + v'y + w'z + r'$ is less or greater than 0. It is then necessary to show that this definition is unique and it is in agreement with the one on p. 8, which is easy to do.

It has thus been seen that all of the Axioms I, II, IV* are satisfied in this space geometry which has been constructed in the manner described above from the Desarguesian set of numbers *D*.

Since Desargues' Theorem is a consequence of Axioms I 1-8, II, IV*, the following becomes evident:

From a **Desarguesian** *set of numbers D it is possible to construct in the manner described above a plane geometry in which the set of the numbers D form the elements of a segment arithmetic in accordance with the one introduced in Section 24 and in which Axioms I 1-3, II, IV*, are satisfied. In such a plane geometry Desargues' Theorem then always holds.*

This result is the converse of the one arrived at in Section 28 and which can be formulated as follows:

Into a plane geometry in which besides Axioms I, 1-3, II, IV* **Desargues' Theorem** *also holds it is possible to introduce a segment arithmetic in accordance with Section 24. By a suitable ordering definition the elements of this segment arithmetic form then a* **Desarguesian** *set of numbers.*

§ 30. The Significance of Desargues' Theorem

If Axioms I, 1-3, II, IV*, are satisfied in a plane geometry in which Desargues' Theorem also holds then by the latter theorem it is always possible to introduce into this geometry a segment arithmetic in which rules 1-11, 13-16 of Section 13 are valid. Let the totality of these segments be still considered as a set of complex numbers and from them let a space geometry be constructed according to the developments of Section 29, in which Axioms I, II, IV* are valid.

Considering in this space geometry only the points $(x, y, 0)$ and those lines on which only such points lie one obtains a plane geometry. Taking into account Theorem 55 deduced in Section 27 it becomes clear that this plane geometry must coincide with the plane geometry assumed at the beginning, i.e., the elements of both geometries can be made to correspond in one-to-one, and onto, manner, and thus preserve their operations and their order. Hence one obtains the following theorem that should be regarded as the final goal of all developments of this chapter:

THEOREM 56. *Let the Axioms* I, 1-3, II, IV*, *be satisfied in a plane geometry. Then the validity of Desargues' Theorem is a necessary and sufficient condition that this plane geometry could be regarded as part of a space geometry in which Axioms* I, II, IV* *hold.*

Desargues' Theorem can thus be characterized for plane geometry as being, so to speak, the result of eliminating the space axioms.

The results obtained also put one into a position to realize that every space geometry in which Axioms I, II, IV* are satisfied can always be regarded as a part of a "geometry of any number of dimensions." By a geometry of any number of dimensions is to be understood a collection of points, lines, planes and other elements for which the corresponding extended axioms of incidence, order, as well as the axiom of parallels are satisfied.

CHAPTER VI

PASCAL'S THEOREM

§ 31. Two Theorems on the Provability of Pascal's Theorem

As has already been noted Desargues' Theorem can be proved from Axioms I, II, IV*, i.e., by essentially using the space axioms but without invoking the congruence axioms. In Section 23 I have shown that its proof without the space axioms of group I and without the congruence axioms III is impossible even if the use of the continuity axiom is allowed.

In Section 14 Pascal's Theorem (Theorem 40), and in Section 22 also Desargues' Theorem, were deduced from Axioms I, 1-3, II-IV, and hence, without the space axioms and by essentially using the congruence axioms. The question arises **whether Pascal's Theorem can also be proved without the use of the congruence axioms** but by invoking the space incidence axioms. The investigations will show that in this respect Pascal's Theorem is completely different from Desargues' Theorem as the **admission or the exclusion of Archimedes' Axiom** in the proof of Pascal's Theorem has a decisive effect on its validity. Since the congruence axioms are generally not assumed in this chapter it is necessary to formulate the Archimedean Axiom in the following form:

V,1* (**Archimedean Axiom** of segment arithmetic). *Let a segment a and two points A and B be given on a line g. Then it is possible to define a number of points* $A_1, A_2, \ldots, A_{n-1}\ A_n$ *such that B lies between A and* A_n *and the segment* $AA_1, A_1A_2, \ldots, A_{n-1}A_n$ *are equal to segment a in the sense of the segment arithmetic which, according to Section 24, can be introduced on the line g on the basis of axioms* I, II, IV* *and Desargues' Theorem.*

The essential results of these investigations can be combined in the following two theorems:

THEOREM 57. *Pascal's Theorem* (Theorem 40) *can be proved with Axioms* I, II, IV*, V, 1*, *i.e., without the congruence axioms but with the aid of Archimedes' Axiom.*

THEOREM 58. *Pascal's Theorem* (Theorem 40) **cannot** *be proved with Axioms* I, II, IV*, *i.e., without the congruence axioms as well as without the Archimedean Axiom.*

In the formulations of both of these theorems it is also possible, by the general Theorem 56, to replace the space axioms I, 4-8 by the plane geometry requirement that Desargues' Theorem (Theorem 53) holds.

§ 32. The Commutative Law of Multiplication in the Archimedean Number Set

The proofs of Theorem 57 and 58 rest essentially on certain mutual relations which exist for the rules of operation and the fundamental properties of ordinary arithmetic, and whose knowledge also appears to be of interest in itself. The following two theorems will now be formulated:

THEOREM 59. *The commutative law of multiplication in an Archimedean number set is a necessary consequence of the other rules of operation, i.e., if a number set has properties* I-II, 13-17 *enumerated in Section* 13 *then it necessarily follows that it also satisfies Formula* 12.

PROOF. First note that if a is any number of the number set and

$$n = 1 + 1 + \ldots + 1$$

is a positive integral rational number then the commutative law of multiplication holds for a and n, namely,

$$an = a(1 + 1 + \ldots + 1) = a \cdot 1 + a \cdot 1 + \ldots + a \cdot 1$$
$$= a + a + \ldots + a$$

and

$$na = (1 + 1 + \ldots + 1)a = 1 \cdot a + 1 \cdot a + \ldots + 1 \cdot a$$
$$= a + a + \ldots + a.$$

Now contrary to the hypothesis, let a and b be two numbers of the number set for which the commutative law of multiplication does not hold. It is clear that it is permissible to make the assumption that

$$a > 0,\ b > 0,\ ab - ba > 0.$$

In view of Requirement 5 of Section 13 there exists a number c (> 0) such that

$$(a + b + 1)c = ab - ba.$$

Finally choose a number d that satisfies the inequalities

$$d > 0,\ d < 1,\ d < c$$

simultaneously and denote by m and n two nonnegative integral rational numbers for which

$$md < a \leqq (m+1)d$$

and

$$nd < b \leqq (n+1)d,$$

respectively. The existence of the numbers m and n is a direct consequence of the Archimedean Theorem (Theorem 17 of Section 13). Referring to the remark made at the outset of this proof one obtains from the last two inequalities by multiplication

$$ab \leqq mnd^2 + (m+n+1)d^2,$$
$$ba > mnd^2,$$

and so by subtraction

$$ab - ba < (m+n+1)d^2.$$

Now

$$md < a, \; nd < b, \; d < 1$$

and hence

$$(m+n+1)d < a+b+1,$$

i.e.,

$$ab - ba < (a+b+1)d$$

and since $d < c$

$$ab - ba < (a+b+1)c.$$

This inequality contradicts the value determined for the number c and the proof of Theorem 59 is thus complete.

§ 33. The Commutative Law of Multiplication in a Non-Archimedean Number Set

THEOREM 60. *The commutative law of multiplication in a non-Archimedean number set is* **not** *a necessary consequence of the other rules of operation, i.e., there exists a number set which has the properties* 1-11, 13-16 *enumerated in Section* 13, *a Desarguesian number set according to Section* 28, *which does not have the commutative law of multiplication* (12).

PROOF. Let t be a parameter and T any expression in a finite or an infinite number of terms of the form

$$T = r_0 t^n + r_1 t^{n+1} + r_2 t^{n+2} + r_3 t^{n+3} + \dots$$

wherein r_0 ($\neq 0$), $r_1, r_2, \dots$ are any rational numbers and n is any integral rational number. Let 0 be adjoined to the range of these expressions T. Two expressions in the form of T will be said to be equal if all numbers $n, r_0, r_1, r_2, \dots$ in them coincide in pairs. Furthermore let s be another parameter and S any expression in a finite or an infinite number of terms of the form

$$S = s^m T_0 + s^{m+1} T_1 + s^{m+2} T_2 + \dots$$

wherein T_0 ($\neq 0$), $T_1, T_2, \dots$ are any expression in the form of T, and let m again be any integral rational number. Consider as a number set $\Omega(s, t)$ the totality of all expressions in the form of S to which the number 0 is adjoined and in which the rules of operation are adopted as follows:

Operate on the parameters s and t themselves according to Rules 7-1 of Section 13 whereas instead of Rule 12 use the formula

$$(1) \qquad ts = 2st.$$

It is easy to convince oneself that this definition is not contradictory.

If S', S'' are any two expressions in the form of S,

$$S' = s^{m'} T_0' + s^{m'+1} T_1' + s^{m'+2} T_2' + \dots,$$

$$S'' = s^{m''} T_0'' + s^{m''+1} T_1'' + s^{m''+2} T_2'' + \dots,$$

then it is clear that by term-by-term addition it is possible to form a uniquely determined new expression $S' + S''$ in the form of S. This expression $S' + S''$ is called the sum of the numbers represented by S' and S''.

By the usual and formal term-by-term multiplication of the two expressions S', S'' one arrives at an expression of the form

$$S'S'' = s^{m'}T_0's^{m''}T_0'' + (s^{m'}T_0's^{m''+1}T_1'' + s^{m'+1}T_1's^{m''}T_0'')$$
$$+ (s^{m'}T_0's^{m''+2}T_2'' + s^{m'+1}T_1's^{m''+1}T_1'' + s^{m'+2}T_2's^{m''}T_0'')$$
$$+ \dots .$$

It is clear that this expression with the use of formula (1) becomes a uniquely determined expression in the form of S. The latter is called the product of the numbers represented by S' and S''.

In this method of operation the validity of the rules of operation 1-4 and 6-11 of Section 13 becomes immediately apparent. The validity of Rule 5 of Section 13 is also not difficult to establish. To show this assume that

$$S' = s^{m'}T_0' + s^{m'+1}T_1' + s^{m'+2}T_2' + \dots$$

and

$$S'''' = s^{m''''}T_0'''' + s^{m''''+1}T_1'''' + s^{m''''+2}T_2'''' + \dots$$

are two expressions in the form of S, and suppose that by definition the first coefficient r_0' of T_0' is different from 0. By equating like powers of s on both sides of the equation

$$S'S'' = S'''' \tag{2}$$

one finds in a uniquely determined manner an integral number m'' first as an exponent and then in succession some expressions

$$T_0'', T_1'', T_2'', \dots$$

such that by using formula (1) the expression

$$S'' = s^{m''}T_0'' + s^{m''+1}T_1'' + s^{m''+2}T_2'' + \dots$$

satisfies equation (2). A corresponding result holds for the equation

$$S'''S' = S''''.$$

The proof is thus complete.

Finally in order to make it possible to order the numbers of the number set Ω (s, t) make the following definitions: A number of the set shall be less or greater than 0 according as the first coefficient r_0 of T_0, in the expression of S which it represents, is less or greater than 0. Given any two numbers a and b of the number set then $a < b$ or $a > b$ according as $a - b < 0$ or $a - b > 0$. It is immediately clear that according to these definitions Rules 13-16 of Section 13 are also valid, i.e., Ω (s, t) is a Desarguesian number set (cf. Section 28).

As formula (1) shows, Rule 12 of Section 13 is **not** valid for the number set Ω (s, t). The validity of Theorem 60 is thus completely established.

In agreement with Theorem 59, the Archimedean Theorem (Theorem 17 of Section 13) does not hold in the number set Ω (s, t) formulated above.

§ 34. Proof of the Two Theorems about Pascal's Theorem (Non-Pascalian Geometry)

If in a space geometry Axioms I, II, IV* are satisfied then Desargues' Theorem (Theorem 53) also holds. Therefore, by the last theorem of Section 28, it is possible to introduce on every pair of intersecting lines in such a geometry a segment arithmetic in which Rules 1-11, 13-16 of Section 13 hold. If the Archimedean Axiom V,1* is assumed in this geometry then it is clear that the Archimedean Theorem (Theorem 17 of Section 13) will hold for the segment arithmetic and hence, by Theorem 59, it will also follow the commutative law of multiplication. From the accompanying figure it becomes immediately clear that the commutative law of multiplication is none other than Pascal's Theorem for the two axes. The validity of Theorem 57 is thus established.

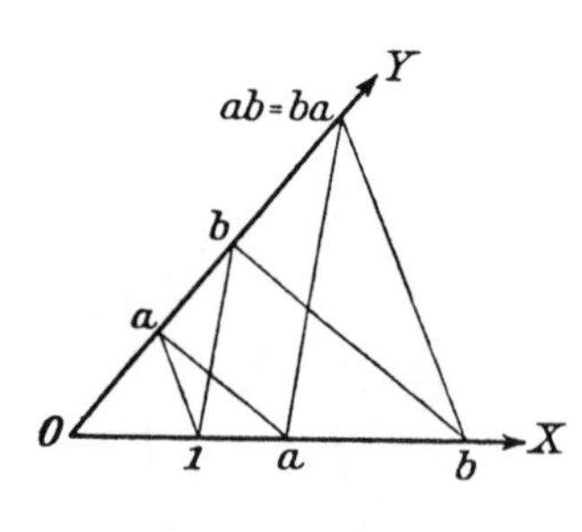

In order to prove Theorem 58 consider the Desarguesian number set Ω (s, t) formulated in Section 33 and with its aid construct in the manner described in Section 29 a space geometry in which Axioms I, II, IV* are satisfied. In this geometry, however, Pascal's Theorem does not hold as the Desarguesian number set Ω (s, t) does not have the commutative law of multiplication. The "*non-Pascalian*" geometry constructed in this manner is, in accordance with Theorem 57 proved previously, necessarily at the same time also a "*non-Archimedean*" geometry.

It is clear that with the assumptions made, it is impossible to prove Pascal's Theorem even if the space geometry is regarded as part of a geometry of any number of dimensions, in which besides the points, lines and planes, there also exist other elements, and for which a corresponding set of axioms of incidence and order as well as the axiom of parallels are assumed.

§ 35. Proof of Any Point of Intersection Theorem by Means of Pascal's Theorem

The following important result is proved next:

THEOREM 61. Desargues' Theorem (Theorem 53) can be proved from Pascal's Theorem (Theorem 40) with the aid of Axioms I, 1-3, II, IV* alone and hence without the use of the congruence and the continuity axioms.

PROOF.[1] It is clear that both assertions of Theorem 53 follow directly from each other. It is therefore sufficient, for example, to prove the second statement of Theorem 53. The proof will be made under some auxiliary hypotheses.

Let two triangles ABC and $A'B'C'$ be positioned so that the lines joining corresponding vertices pass through the point O, AB is parallel to $A'B'$ and AC is parallel to $A'C'$. Assume further that either the lines OB' and $A'C'$ are parallel or the lines OC' and $A'B'$ are parallel. Draw the parallel to OB' through A which is met by the line $A'C'$ at a point L and by the line OC' at a point M. Furthermore, let the line LB' be parallel either to OA or to OC. The lines AB and LB' are certainly not parallel, i.e., they meet at a point N. Join it with M and O.

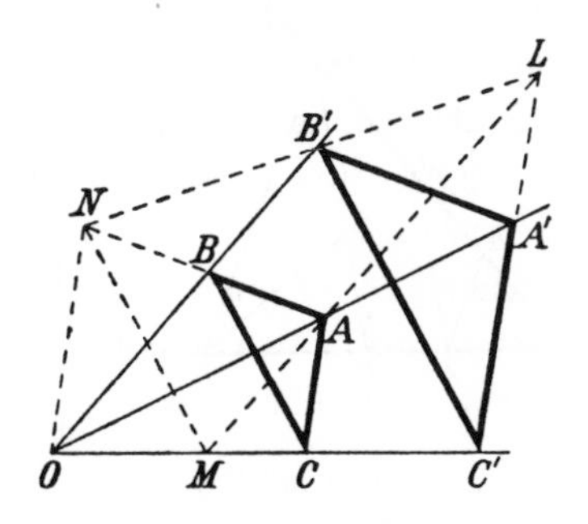

[1] Theorem 61 was proved by G. Hessenberg ("Beweis des Desarguesschen Satzes aus dem Pascalschen," *Math. Ann.*, Vol. 61), in the manner shown below.

According to the construction Pascal's Theorem can be applied to the configuration $ONALA'B'$ and it can be seen that ON is parallel to $A'L$ and hence is also parallel to CA. Furthermore, Pascal's Theorem can also be applied to the configuration $ONMACB$ and $ONMLC'B'$ and shows that MN is parallel to CB as well as to $C'B'$. Hence, the sides CB and $C'B'$ are parallel.

The auxiliary hypotheses made for the proof can be dropped now one by one. The proof of these possibilities will be omitted here.

Now let a plane geometry be given in which besides Axioms I, 1-3, II, IV* Pascal's Theorem holds as well. Theorem 61 shows that Desargues' Theorem also holds in this geometry. Hence, it is possible to introduce into this geometry a segment arithmetic in accordance with Section 24. According to Section 34 Pascal's Theorem holds in this segment arithmetic as does also the commutative law of multiplication, i.e., all Rules of Operation 1-12 of Section 13 are valid in it.

Denoting a figure which corresponds to the content of Pascal's or Desargues' Theorem as a Pascalian or Desarguesian configuration, respectively, it is possible to combine the results of Sections 24-26 and 36 as follows: Every application of the rules of operation (Theorems 1-12 of Section 13) in this segment arithmetic turns out to be a combination of a finite number of Pascalian and Desarguesian configurations. Since according to the proof of Theorem 61, Desarguesian configurations can be represented as a combination of Pascalian configurations by constructing suitable auxiliary points and lines, every application of these rules of operation in this segment arithmetic turns out to be a combination of a finite number of Pascalian configurations.

By Section 27 and with the commutative law of multiplication it is possible to represent a point in this segment arithmetic by a pair of real numbers (x, y) and a line by a ratio of three real numbers $(u : v : w)$ in which the first two do not vanish simultaneously. The common locus of a point and a line is characterized by the equation

$$ux + vy + w = 0$$

and the parallelism of two lines $(u : v : w)$ and $(u' : v' : w')$ by the proportion

$$u : v = u' : v'.$$

Now let a pure point of intersection theorem be proposed in the geometry specified in this manner. A pure point of intersection theorem is to be understood here as a theorem that contains an assertion about the common locus of points and lines and the parallelism of lines without the use of other relations such as congruence and perpendicularity. Every such pure point of intersection theorem in a plane geometry can be put in the following form:

Choose an arbitrary set of a finite number of points and lines. Then draw in a prescribed manner any parallels to some of these lines. Choose any points on some of the lines and draw any lines through some of these points. Then, if connecting lines, points of intersection and parallels are constructed through the points existing already in the prescribed manner, a definite set of finitely many lines is eventually reached, about which the theorem asserts that they either pass through the same point or are parallel.

Consider now the coordinates of the completely arbitrarily chosen points and lines as parameters $p_1, \ldots, p_n$. Some coordinates of the less arbitrarily chosen points and lines can be considered as additional parameters $p_{n+1}, \ldots, p_r$. These elements are thus represented by the parameters $p_1, \ldots, p_r$. The coordinates of all connecting lines, points of intersection and parallels which can be constructed now, become then expressions $A(p_1, \ldots, p_r)$ which are rationally dependent on these parameters and the assertion of the proposed point of intersection theorem is represented by the statement that some expressions of the type will yield equal values for equal values of the parameters, i.e., the point of intersection theorem will state that certain expressions $R(p_1, \ldots, p_r)$ depending rationally on certain parameters $p_1, \ldots, p_r$ vanish as soon as some elements from the segment arithmetic introduced into the proposed geometry, are substituted for these parameters. Since the domain of these elements is infinite one concludes by a well-known theorem of algebra that the expressions $R(p_1, \ldots, p_r)$ must vanish **identically on the basis of the Rules of Operation** 1-12 of Section 13. However, in order to prove that the expressions $R(p_1, \ldots, p_r)$ vanish identically in the segment arithmetic it is sufficient now, as has already been shown for the application of the rules of operation, to apply Pascal's Theorem, and the following is thus seen:

THEOREM 62. *Every pure point of intersection theorem that holds in a plane geometry in which Axioms* I, 1-3, II, IV* *and Pascal's Theorem are valid takes, through the construction of suitable auxiliary* *points and lines, the form of a combination of finite number of Pascalian configurations.*

In proving the point of intersection theorem with the aid of Pascal's Theorem it is then no longer necessary to revert to the congruence and continuity axioms.

CHAPTER VII

GEOMETRIC CONSTRUCTIONS BASED ON AXIOMS I-IV

§ 36. Geometric Constructions with Ruler and Compass

Let a space geometry be given in which Axioms I-IV hold. For the sake of simplicity only a **plane** geometry which is contained in this space geometry will be considered in this chapter. The question of which elementary construction problems (given suitable practical aids) can necessarily be carried out in such a geometry will be investigated.

The following problems are always solvable on the basis of Axioms I, II, IV:

PROBLEM 1. Join two points by a line and find the point of intersection of two lines in case they are not parallel.

The construction of segments and angles is possible on the basis of the congruence axioms III, i.e., the following problems can be solved in the given geometry:

PROBLEM 2. Construct a given segment on a given line from a point in a given direction.

PROBLEM 3. Construct a given angle on a given line at a given point on a given side or construct a line that intersects a given line at a given point at a given angle.

It can be seen that by assuming Axioms I-IV only those construction problems which can be reduced to Problems 1-3, given above, are solvable.

The following two problems will be adjoined to fundamental Problems 1-3:

PROBLEM 4. Draw a parallel to a line through a given point.

PROBLEM 5. Construct a perpendicular to a given line.

It can be seen at once that both of these problems can be solved in different ways by means of Problems 1-3.

To carry out the construction of Problem 1 one requires a **ruler**. In order to solve Problems 2-5 it is sufficient, as will subsequently be shown, to use in addition to the ruler, a scale, an instrument with which to lay off a single[1] fixed segment, like the unit segment. Hence the following result can be arrived at:

THEOREM 63. *Those geometric construction problems which are solvable on the basis of Axioms* I-IV *can necessarily be carried out with a ruler and a scale.*

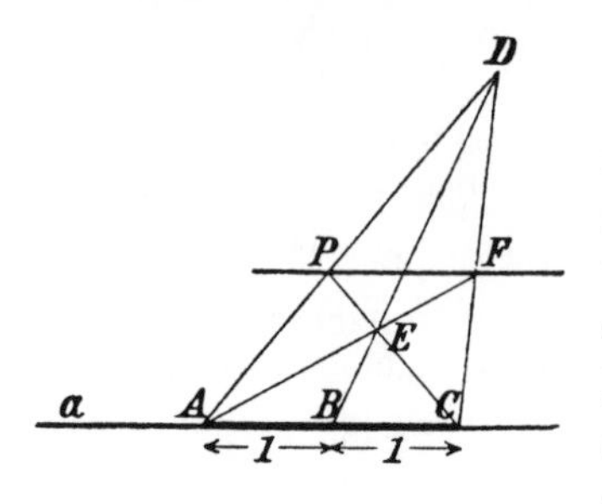

PROOF. In order to carry out the construction of Problem 4 join the given point *P* with any point *A* of the given line *a* and from *A* lay off with the scale the unit segment twice contiguously on *a*, say to *B* and *C*. Now let *D* be any point on *AP* that is distinct from *A* and *P* and for which *BD* is not parallel to *PC*. Then *CP* and *BD* meet at a point *E* and *AE* and *CD* meet at a point *F*. According to Steiner *PF* is the desired parallel to *a*.

Problem 5 is solved in the following way: Let *A* be any point of the given line. Lay off with the scale on both sides of *A* the unit segments *AB* and *AC* and then determine on any two lines through *A* the points *E* and *D* such that the segments *AD* and *AE* are also equal to the unit segment. The lines *BD* and *CE* intersect at a point *F*, the lines *BE* and *CD* intersect at a point *H* and *FH* is the desired perpendicular. In fact, the angles ∡ *BDC* and ∡ *BEC*, as angles inscribed in a semicircle whose diameter is *BC*, are right angles. Hence by the point of intersection theorem of the altitudes of a triangle, applied to the triangle *BCF*, *FH* is also a perpendicular to *BC*.

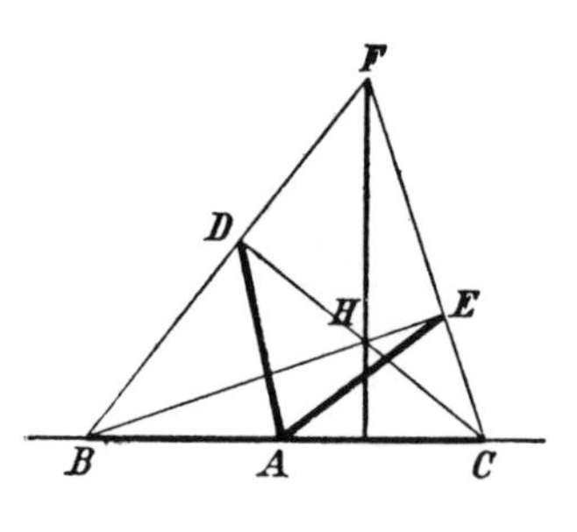

On the basis of Problems 4 and 5 it is always possible to drop a perpendicular to a given line *a* from a point *D* not on the line, or to erect on it a perpendicular at a point *A* lying on the line.

[1] J. Kürschák has observed that the requirement here of constructing a single segment is sufficient. Cf. his note "Das Streckenabtragen," *Math. Ann.*, Vol. 55 (1902).

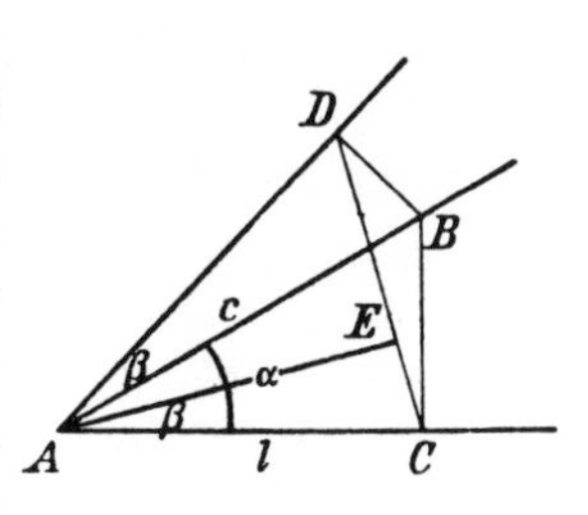

It is easy now to solve Problem 3 with only a ruler and a scale. Let the following method, which requires only drawing parallels and dropping perpendiculars, be adopted: Let β be the angle to be constructed and A its vertex. Draw the line l through A parallel to the given line on which the given angle β is to be constructed. From any point B on one side of β drop perpendiculars to the other side of the angle β and to l. Let the feet of these perpendiculars be D and C. They are distinct and A does not lie on CD. Hence it is possible to drop a perpendicular from A to CD. Let its foot be E. According to the proof given on page 47 $\measuredangle\ CAE = \beta$. If B is chosen on the other side of the given angle then E lies on the other side of l. Draw then a parallel to AE through the given point on the given line. Problem 3 is thus solved.

Finally in order to carry out the construction on Problem 2 the following simple method given by J. Kürschák will be used: Let AB be the segment to be constructed and P the given point on the given line l.

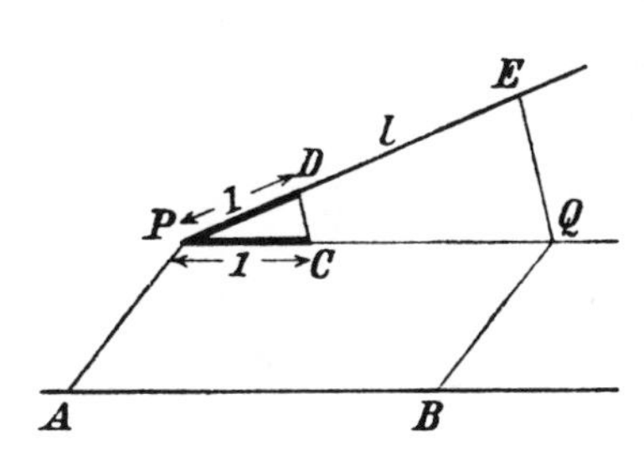

Draw the parallel to AB through P and with the scale lay off on it, on the side of AP on which B lies, the unit segment from the point P, say to C. Furthermore, construct the unit segment on l from P to D on the given side. Let the parallel to AP drawn through B meet PC at Q and the parallel to CD drawn through Q meet l at E. Then $PE = AB$. In case l coincides with PQ and Q does not lie on the given side the construction can be continued in a simple manner.

It has been shown that Problems 1-5 are all solvable with ruler and scale and consequently the proof of Theorem 63 is complete.

§ 37. Criterion for the Possibility of Geometric Constructions with Ruler and Compass

Aside from the elementary geometric problems treated in Section 36 there is a long series of other problems whose solutions require only drawing lines and construction of segments. In order to survey the domain of all problems which can be solved in this manner an orthogonal coordinate system is introduced in the subsequent discussion and the coordinates of the points are considered as real numbers or functions of some arbitrary parameters in the usual way. In order to answer the question about the totality of all constructable points the following approach is taken:

Let there be given a set of fixed points. From the coordinates of these points form a rational domain R. It contains some real numbers and some arbitrary parameters p. Consider now the collection of all points which can be constructed from the given set of points by drawing lines and construction of segments. Let the domain that is formed from the coordinates of these points be denoted by Ω (R). It contains some real numbers and functions of the arbitrary parameter p.

The discussions of Section 17 show that drawing lines and parallels is analytically equivalent to the operation of addition, multiplication, subtraction and division of segments. Furthermore, the well-known formula for a rotation given in Section 9 shows that construction of segments on any line requires no analytic operation other than the extraction of square roots from the sum of two squares whose bases have already been constucted. Conversely, the square root of the sum of two squared segments can always be laid off through segment construction on the basis of Pythagoras' theorem and with the aid of a right triangle.

It follows from these discussions that the domain Ω (R) contains all those and only those real numbers and functions of the parameter p which arise from a finite number of applications of the five arithmetic operations to the numbers and parameters in R, namely, the four elementary arithmetic operations and a fifth operation, the extraction of a square root from the sum of two squares. The result is expressed as follows:

THEOREM 64. A geometric construction problem is solvable by drawing lines and construction of segments, i.e., with ruler and scale if and only if in the analytic treatment of the problem the coordinates of the desired points are functions of the coordinates of the given points whose formation requires only rational operations and the operation of extracting square roots from the sum of two squares and requires the application of these five operations only a finite number of times.

It becomes immediately apparent from this theorem that not every problem solvable with a compass is also solvable with a ruler and scale alone. To show this consider the geometry which was constructed in Section 9 with the aid of the algebraic number field Ω. In this geometry there exist only segments which are constructable with ruler and scale, namely, the segments determined by the numbers of the field Ω.

Now if ω is any number in Ω, it can be seen from the definition of the field Ω, that every number conjugate to ω must also be in Ω. Since it is clear that all numbers of Ω are real, it follows that the field can contain only such numbers whose conjugates are also real, i.e., all numbers of the field Ω are real.

Let the problem of constructing a right triangle with a hypotenuse of 1 and with a leg of $|\sqrt{2}|-1$ be given. The algebraic number $\sqrt{2|\sqrt{2}|-2}$, which expresses the numerical value of the other leg is not in the field Ω since its conjugate $\sqrt{-2|\sqrt{2}|-2}$ is imaginary. Hence the given problem is unsolvable in the assumed geometry and thus cannot be solved at all with ruler and scale, although the construction with a compass can be performed immediately.

This analysis is also reversible, i.e., the following holds: Every real number obtained from rational numbers through real square roots, is in the field Ω. Hence, every segment determined by such a number is constructable with ruler and scale. The proof of this theorem will be obtained from more general considerations. It is indeed possible to find a criterion for the solvability of a geometric construction problem, with ruler and compass, which allows one to determine immediately from the analytical nature of the problem and its solution whether the construction is also performable with only ruler and scale. This is given by the following theorem:

THEOREM 65. *Let a geometric construction problem be given such that in its analytic treatment the coordinates of the desired points can be found from the coordinates of the given points merely by rational operations and the extraction of square roots. Let n be the smallest number of the square roots that is sufficient to compute the coordinates of the points. Should the given construction problem be also performable only by drawing lines and constructing segments, then by including the point at infinity, it is necessary and sufficient that the geometric problem have exactly 2^n real solutions for* **all** *positions of the given points,*[1] i.e., *for all values of the arbitrary parameters that occur in the coordinates of the given points.*

[1] See Suppl. IV, 2.

On the basis of the considerations given at the beginning of this section the necessity of the formulated criterion follows immediately. The assertion that this criterion is also sufficient, is equivalent to the following arithmetical theorem:

THEOREM 66. *Let a function* $f(p_1, \ldots, p_n)$ *of the parameters* $p_1, \ldots, p_n$ *be formed by rational operations and extraction of square roots. If for* **every** *set of real values of the parameters it represents a* **real** *number then it is in the field* $\Omega(R)$ *which is obtained by the elementary arithmetic operations and extraction of square roots* **from a sum of two squares** *starting with* $1, p_1, \ldots, p_n$.

Let it be remarked that the restriction to sums of squares of **two summands** can be dropped from the definition of $\Omega(R)$. In fact the formulas

$$\sqrt{a^2 + b^2 + c^2} = \sqrt{(\sqrt{a^2 + b^2})^2 + c^2},$$

$$\sqrt{a^2 + b^2 + c^2 + d^2} = \sqrt{(\sqrt{a^2 + b^2 + c^2})^2 + d^2},$$

$$\cdots\cdots\cdots\cdots\cdots\cdots$$

show that the extraction of square roots from a sum of any number of squares can always be reduced to the repeated extraction of square roots from the sum of two squares.

According to this, by considering the rational fields which arise from the step-by-step adjunction of the innermost square roots in the construction of the function $f(p_1, \ldots, p_n)$ it is sufficient to prove that the radicand can be represented in the preceding rational field as a sum of squares. The proof of this will be based on the following algebraic theorem:

THEOREM 67. *Every rational function* $\rho(p_1, \ldots, p_n)$ *with rational coefficients which never assumes negative values for real values of the parameters can be represented as a sum of squares of rational functions of the variables* $p_1, \ldots, p_n$ *with rational coefficients.*[1]

This theorem will be given the following form:

THEOREM 68. *In the rational field generated by* $1, p_1, \ldots, p_n$ *every function which for no set of real values of the variables assumes negative values is a sum of squares.*

[1] This problem was first treated by me for **one** variable where E. Landau proved the theorem completely for one variable with the use of very simple and elementary tools., *Math. Ann.*, Vol. 57 (1903). The complete proof has been recently given by Artin, *Hamburger Abhandlungen*, Vol. 5 (1927).

Let $f(p_1, \ldots, p_n)$ be a function with the properties given in Theorem 66. Extend the last assertion to the fields which are obtained from step-by-step adjunctions of these square roots which are required for the construction of the function f. In these fields, it is true that every nonnegative function, together with all its conjugates, can be represented as a sum of squares of functions from the field under consideration.

The proof will be given by induction. Consider first a field that arises from R by the adjunction of an innermost square root of the function. The radicand of this square root is a **rational** function $f_1(p_1, \ldots, p_n)$. Let $f_2(p_1, \ldots, p_n)$ be a function from the field $(R; \sqrt{f_1})$ that arises from the adjunction, which, together with all its conjugates, never assumes negative values, and which also never vanishes identically. It has the form $a + b\sqrt{f_1}$, where a and b, as well as f_1, are rational functions. From the hypotheses made about f_2 it follows that the sum φ, and the product ψ, of the functions $a + b\sqrt{f_1}$, $a - b\sqrt{f_1}$ never assume negative values. The functions

$$\varphi = 2a, \quad \psi = a^2 - b^2 f_1$$

are therefore rational, and by Theorem 68 can be represented as sums of squares of functions from R. Besides, φ cannot vanish identically.

From the equation

$$f_2^2 - \varphi f_2 + \psi = 0$$

satisfying f_2, one obtains

$$f_2 = \frac{f_2^2 + \psi}{\varphi} = \left(\frac{f_2}{\varphi}\right)^2 \cdot \varphi + \frac{\varphi\psi}{\varphi^2}.$$

According to the descriptions of φ and ψ, f_2 can be represented as a sum of squares of functions from the field $(R, \sqrt{f_1})$. The result thus obtained for the field $(R, \sqrt{f_1})$ corresponds to Theorem 68, which holds in the field R. By repeating the procedure applied above for further adjunctions one arrives finally at the result that in every one of the fields which are obtained by the construction of the function f, every nonnegative function, together with all its conjugates, is a sum of squares of functions from the field under consideration. Consider any square root occuring in f. It, together with its conjugates, is always real.

Hence, its radicand, together with all its conjugates, are nonnegative functions in the field in which it can be represented, and thus can be represented in that field as a sum of squares. Theorem 65 is thus proved The criterion given in Theorem 65 is therefore sufficient.

The regular polygons which are constructable with a compass can be taken as an example of the application of Theorem 65. In this case an arbitrary parameter p does not occur. The expressions to be constructed are rather represented by algebraic numbers. It is easy to see that the criterion of Theorem 65 is satisfied, and hence it follows that every regular polygon can be constructed by only drawing straight lines and constructing segments, a result which can also be deduced directly from the theory of cyclotomy.

As far as other well-known construction problems of elementary geometry are concerned, it will only be mentioned here that the problem of Malfatti, but not the contact problem of Apollonius, is solvable with only ruler and scale.[1]

CONCLUSION

The present treatment is a critical investigation of the principles of geometry. In this investigation the ground rule was to discuss every question that arises in such a way so as to find out at the same time whether it can be answered in a specified way with some limited means. This ground rule seems to me to contain a general and natural guide-line. In fact, if in the course of mathematical investigations, a problem is encountered, or a theorem is conjectured, the drive for knowledge is then satisfied only if either the complete solution of the problem and the rigorous proof of the theorem are successfully demonstrated or the basis for the impossibility of success and hence the inevitability of failure are clearly seen.

The **impossibility** of certain solutions and problems thus plays a prominant role in modern mathematics, and the drive to answer questions of this type was oftentimes the cause for the discovery of new and fruitful areas of investigation. Recall only Abel's proof of the

[1] For other geometric construction problems with ruler and scale, cf. M. Feldblum, *Über elementargeometrische Konstruktionen*, Inaugural Dissertation (Göttingen, 1899).

impossibility of solving the fifth degreee equation by radicals, the realization of the impossibility of proving the axiom of parallels, and Hermites' and Lendemann's theorem on the impossibility of constructing the numbers e and π algebraically.

The ground rule according to which the principles of the possibility of a proof should be discussed at all is very intimately connected with the requirement for the "purity" of the methods of proof which has been championed by many mathematicians with great emphasis. This requirement is basically none other than a subjective form of the ground rule followed here. In fact, the present geometric investigation seeks to uncover which axioms, hypotheses or aids are necessary for the proof of a fact in elementary geometry, and in every decision the question as to which method of proof is to be preferred, from the adopted point of view, remains open.

APPENDIX I

THE STRAIGHT LINE AS THE SHORTEST DISTANCE BETWEEN TWO POINTS[1]

(from *Math. Ann.*, Vol. 46)
(from a letter to F. Klein)

If points, lines, and planes are taken as elements, then the following axioms can serve for the foundations of geometry:

1. **The axioms of the elements' mutual relations**; briefly summarized, these can be stated as follows:

Any two points A and B determine a line a. – Any three points A, B, C, which do not lie on a line, determine a plane α *– If two points A, B, of a line a, lie in a plane then the line a lies completely in the plane* α *– If two planes* α, β *have a point A in common, then they have at least another point B in common. – On every line there exist at least two points, in every plane there exist three points that do not lie on a line, and in space there exist at least four points which do not lie in a plane.*

2. **The axioms of segments and sequences of points on a line.** These axioms were first introduced by M. Pasch[2] and then investigated by him systematically. They are essentially the following:

Between two points A and B of a line there exists at least a third point C of the line. – Among three points of a line there exists one and only one which lies between the other two. – If A and B lie on a line a then there exists a point C of the same line a such that B lies between A and C. – Any four points A_1, A_2, A_3, A_4 *of a line a can be ordered so that* A_i *lies between* A_h *and* A_k *as soon as the index h is less than and k is greater than i. – Every line a that lies in a plane* α *partitions the points of this plane into two regions with the following properties: Every point A of one region together with every point* A' *of the other region determine a segment* AA' *on which lies a point of the line a. However, two points A and B of the same region determine a segment AB which contains no point of the line a.*

[1] Concerning a more general formulation of these problems, compare my lecture delivered at the International Mathematical Congress in Paris in 1900: "Mathematische Probleme," *Göttinger Nachrichten,* No. 4 (1900), as well as G. Hamel's Inaugural Dissertation (Göttingen, 1901), and his article "über die Geometrien, in denen die Geraden die Kürzesten sind," *Math. Ann.*, Vol. 57 (1903).

[2] Cf. *Vorlesungen über neuere Geometrie* (Teubner, 1882).

3. **The axiom of continuity.** To it I give the following form: *If* A_1, A_2, $A_3, \ldots$ *is an infinite sequence of points of a line a and B is another point of a such that* A_i *lies between* A_h *and B as soon as the index h is less than i then there exists a point C with the following property: All points of the infinite sequence* A_2, A_3, $A_4, \ldots$ *lie between* A_1 *and C and if C′ is another point for which this is also true then C lies between* A_1 *and C′.*

The theory of harmonic points can be established with complete rigor with these axioms, and by utilizing it as **F. Lindemann**[1] did, the following theorem is obtained:

To every point it is possible to assign three real numbers x, y, z and to every plane it is possible to assign a linear relation among these three points such that all points for which the three points x, y, z satisfy the linear relation lie in this plane, and conversely, to all points that lie in this plane there correspond points x, y, z which satisfy the linear relation. Furthermore, if x, y, z are taken as the orthogonal coordinates of a point in ordinary Euclidean space, then to the points of the original space there correspond points in the interior of some nowhere concave body of the Euclidean space, and conversely, to all points of the interior of the nowhere concave body there correspond points of the original Euclidean space. *Hence, the original space is mapped onto the interior of a nowhere concave body of the space.*

By a nowhere concave body is to be understood here a body with the property that if two points lying in its interior are joined by a line, the part of the line lying between these two points remains entirely in the interior of the body. I take the liberty of pointing out to you that the nowhere concave bodies occurring here also play an important role in the number theory investigations of H. Minkowski,[2] and that he has found a simple analytic definition for them.

Conversely, if any nowhere concave body is given in Euclidean space, then it defines a definite geometry in which all the mentioned axioms are satisfied. To every point in the interior of the nowhere concave body corresponds a point of that geometry. Every line and plane of the Euclidean space passing through the interior of the body corresponds to a line and plane of the general geometry, respectively.

[1] Cf. Clebsch-Lindemann, *Vorlesungen über Geometrie*, Vol. II, Part 1, pp. 433 f.

[2] Cf. *Geometrie der Zahlen* (Teubner, 1896 and 1910).

To the points lying on the boundary or in the exterior of the nowhere concave body and to the lines and planes of the Euclidean space lying entirely in the exterior of the body there correspond no elements of the general geometry.

The above theorem on the mapping of the points of the general geometry onto the interior of the nowhere concave body in Euclidean space thus expresses a property of the element of the general geometry which is completely equivalent in content to the axioms formulated at the outset.

The concept of the length of a segment AB in the general geometry will now be defined, and to this end, the two points of the Euclidean space which correspond to the points A and B of the original space will also be denoted by A and B. Extend now the line AB in the Euclidean space beyond A and B up to where it meets the boundary of the nowhere concave body at the points X and Y, respectively. Denote in general the distance between any two points P and Q in the Euclidean space by $\overline{PQ}$. Then the real value

$$\widehat{AB} = \log\left\{\frac{\overline{YA}}{\overline{YB}} \cdot \frac{\overline{XB}}{\overline{XA}}\right\}$$

will be called the length of the segment AB in the general geometry. In view of

$$\frac{\overline{YA}}{\overline{YB}} > 1, \qquad \frac{\overline{XB}}{\overline{XA}} > 1$$

length is always a positive quantity.

The properties of length, which necessarily lead to an expression of the type given for $\widehat{AB}$, can easily be enumerated. However, I shall omit them in order not to exhaust your courtesy with this letter.

The formula given for $\widehat{AB}$ shows at the same time how these quantities depend on the shape of the nowhere concave body. Holding the A and B fixed in the interior of the body and varying only the boundary of the body in such a way so that the boundary point X moves toward A and Y approaches the point B it is clear that each of the two quotients

$$\frac{\overline{YA}}{\overline{YB}}, \quad \frac{\overline{XB}}{\overline{XA}}$$

and hence also the value of $\widehat{AB}$, increases.

Let a triangle ABC be given now in the interior of the nowhere concave body. A plane α in it carves out of the body a nowhere concave oval. Suppose further that each of the sides AB, AC, BC of the triangle are extended beyond both end points up to where they intersect the boundary of the oval at the points X and Y, U and V, T and Z. Then construct the connecting lines UZ and TV and extend both of these up to their intersection W. Denote their points of intersection with the line XY by X' and Y'. Considering now the triangle UWT instead of the nowhere concave oval in the plane α *it is* easy to see that the lengths $\overset{\frown}{AC}$ and $\overset{\frown}{BC}$ in the geometry determined by this triangle are the same as in the original geometry while the length of the side AB has increased by the brought-about change. Let the new length of the side AB be denoted by $\overset{\frown}{\overset{\frown}{AB}}$ to distinguish it from the original length $\overset{\frown}{AB}$. Then $\overset{\frown}{\overset{\frown}{AB}} > \overset{\frown}{AB}$.

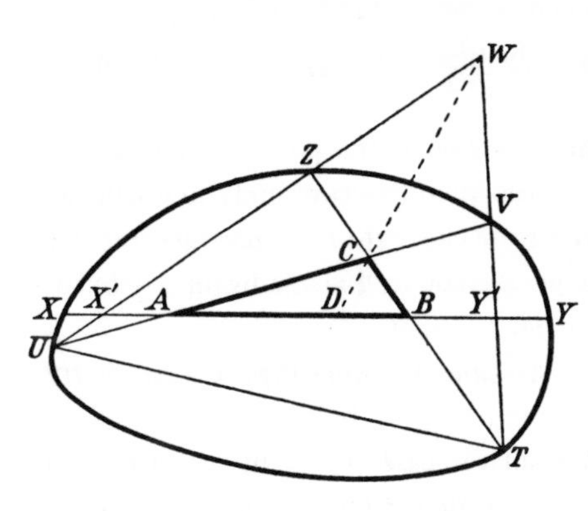

There exists now the simple relation

$$\overset{\frown}{\overset{\frown}{AB}} = \overset{\frown}{AC} + \overset{\frown}{BC}$$

between the sides of the triangle ABC. For the proof connect W with C and extend this line up to the point of intersection D with AB. Then by the well-known cross ratio theorem, in view of the perspective positions of the two sequences of points X', A, D, Y' and U, A, C, V,

$$\frac{\overline{Y'A}}{\overline{Y'D}}\,\frac{\overline{X'D}}{\overline{X'A}} = \frac{\overline{VA}}{\overline{VC}}\,\frac{\overline{UC}}{\overline{UA}},$$

and in view of the perspective positions of the two sequences of points Y', B, D, X' and T, B, C, Z,

$$\frac{\overline{X'B}}{\overline{X'D}}\,\frac{\overline{Y'D}}{\overline{Y'B}} = \frac{\overline{ZB}}{\overline{ZC}}\,\frac{\overline{TC}}{\overline{TB}}.$$

Multiplication of the two equations yields

$$\frac{\overline{Y'A}}{\overline{Y'B}}\,\frac{\overline{X'B}}{\overline{X'A}}=\frac{\overline{VA}}{\overline{VC}}\,\frac{\overline{UC}}{\overline{UA}}\cdot\frac{\overline{ZB}}{\overline{ZC}}\,\frac{\overline{TC}}{\overline{TB}},$$

and this new equation proves my assertion.

From the above investigations you can see that by the given definition of length the following general theorem necessarily holds merely on the basis of the axioms enumerated at the beginning of my letter and by the simplest properties of length.

In every triangle the sum of two sides is greater than or equal to the third side.

It is clear at the same time that the case of equality occurs if and only if the plane α delineates two **rectilinear** segments UZ, and TV, on the boundary of the nowhere concave body. The last condition can also be expressed without the aid of the nowhere concave body. Indeed, if any two lines a and b, which lie in a plane α, and which intersect at a point C, are given in the original geometry, then in general there will exist lines in each of the resulting four plane angular regions in α about C, which intersect neither of the two lines a and b. In particular, however, if there exist no such lines in two oppositely located plane angular regions, then the condition in question is satisfied, and *there exist then triangles in which the sum of two sides is equal to the third.* In this case a path consisting of two rectilinear segments between some points A and B, whose total length is equal to the direct distance between the two points A and B, is possible. It can easily be shown that *all paths between the two points A and B with the same property can be made up of the constructed paths and that the total lengths of other paths are greater.* A closer study of the problem of shortest paths can easily be made, and is of particular interest in the case when a tetrahedron is taken as the boundary of the nowhere concave body.

In conclusion I take the liberty of pointing out that in the preceding development I have always assumed the nowhere concave body to lie in the finite part of space. However, if in the geometry defined by the original axioms there exist a line and a point with the property that through that point only one parallel is possible to that line, then that assumption cannot be justified. It is easy to see what modifications have then to be made to my arguments.

Kleinteich bei Ostseebad Rauschen, August 14, 1894.

APPENDIX II

THE THEOREM ON THE EQUALITY OF THE BASE ANGLES OF AN ISOSCELES TRIANGLE

This appendix, which is a revision of my article **"Über den Satz von der Gleichheit der Basiswinkel"**[1] is concerned with the place of this theorem in **plane Euclidean geometry**.

The following axioms will be assumed here:

I. The **plane** axioms of incidence, i.e., Axioms I, 1-3 (p. 3);

II. The Axioms of order (p. 4-5);

III. The following axioms of congruence:

Axioms III, 1-4 **(pp. 10-13)** in **unchanged** form, the triangle congruence Axiom III, 5 in a restricted form, by requiring its validity only for triangles of the **same orientation**. The orientation of a triangle in a plane geometry is defined, on p. 64, by distinguishing between "**right**" and "**left.**" The definition of a right-hand and a left-hand side of a line shows immediately that one of the sides of any angle is always to be denoted as the right-hand side, and the other as the left-hand side, in a uniquely determined way, namely, so that the right-hand side of the angle lies to the right of the line which is determined by the other side of the angle by position and sense, while the left-hand side of the angle lies to the left of that line which is determined by the other side of the angle by position and sense. The right-hand sides of two angles will be called *equipositional* with respect to these angles. The same will be done for the two left-hand sides.

The restricted form of the triangle congruence axiom will now read as follows:

III, 5*. *If for two triangles ABC and $A'B'C'$ the congruences $AB \equiv A'B'$, $AC \equiv A'C'$ and $\measuredangle BAC \equiv \measuredangle B'A'C'$ are satisfied, then the congruence*

$$\measuredangle ABC \equiv \measuredangle A'B'C'$$

is also satisfied provided that AB and $A'B'$ and **equipositional** *sides of the angles $\measuredangle BAC$ and $\measuredangle B'A'C'$, respectively.*

From the broader form III, 5 of this axiom and from the second part of Axiom III, 4 the "base angle theorem," Theorem 11, (see p. 14), follows immediately. Conversely, the broader form III, 5 can be proved with the aid of Axioms I, II, III, 1-4, III, 5* introduced here, the base angle theorem, and the two following axioms:

[1] *Proceedings of the London Math. Soc.*, Vol. 35.

III, 6. *If the angle* $\measuredangle (h', k')$ *and the angle* $\measuredangle (h'', k'')$ *are congruent to the angle* $\measuredangle (h, k)$ *then the angle* $\measuredangle (h', k')$ *is also congruent to the angle* $\measuredangle (h'', k'')$.

The statement of this axiom was proved with the aid of the broader form III, 5, on p. 18 as Theorem 19.

III, 7. *If two rays c and d which emanate from the vertex of an angle* $\measuredangle (a, b)$ *lie in the interior of this angle then the angle* $\measuredangle (a, b)$ *is not congruent to the angle* $\measuredangle (c, d)$.

The proof of Axiom III, 5 with the aid of the mentioned axioms and the base angle theorem will be omitted here.[1]

IV. The axiom of parallels can be taken here in the weaker form IV (p. 25).

V. The following axioms of continuity:

Archimedes' Axiom V, 1 (p. 26).

(The completeness axiom V, 2 is **not** used here.)

V, 3 (Neighborhood axiom). *Given any segment AB there exists a triangle in whose interior it is impossible to find a segment that is congruent to AB.*

This axiom can be proved with the aid of the broader Form III, 5 of the triangle congruence axiom. The proof rests on the following theorem that can be deduced from Theorems 11 and 23. "The sum of two sides of a triangle is greater than the third side."

The following result whose proof will be omitted is true.[2]

The base angle theorem (Theorem 11), *and hence also the broader form* III, 5 *of the triangle congruence axiom, can be proved from the axioms introduced under* I-V.

The question arises whether it is also possible to prove the broader form of the triangle congruence axiom from the restricted form without the axioms of continuity V, 1, 3. *The following investigation will show that neither can Archimedes' Axiom be dropped, nor can the neighborhood axiom be left out even if the theorems of the theory of proportion are assumed.* In my opinion, the geometries, which I

[1] The observation that instead of a far-reaching axiom by W. Zabel, formerly used in this proof, Axiom III, 7, as presented here, is sufficient, is due to P. Bernays, cf. the footnote on p. 221.

[2] The proof is given by Arnold Schmidt in the article "Die Herleitung der Spiegelung aus der ebenen Bewegung," *Math. Ann.*, Vol. 109 (1934). See Theorem 9 (pp. 561-62) and Set No. 6 in the list at the end.

construct subsequently for this purpose shed new light on the logical connection of the theorem of the isosceles triangle with the other elementary theorems of plane geometry, in particular, with the theory of area.

Let t be a parameter and α any expression in a finite or infinite number of items of the form

$$\alpha = a_0 t^n + a_1 t^{n+1} + a_2 t^{n+2} + \ldots .$$

In it let a_0 ($\neq 0$), a_1, a_2 ... denote arbitrary real numbers, and let n be any integral number (greater than, less than or equal to 0). Consider the totality of all numbers of the form α to which 0 is adjoined as a number of the set T in the sense of Section 13 by satisfying the following conventions: Numbers of the set T are added, subtracted, multiplied and divided as if they were ordinary absolutely convergent power series, arranged according to increasing powers of the variable t. The resulting sums, differences, products and quotients are again expressions in the form of α, and hence are numbers of the set T. Call a number α in T less than or greater than 0 according as in the given expression for α, the first coefficient a_0 is less than or greater than 0. If any two numbers α, β of the number set T are given, then $\alpha < \beta$ or $\alpha > \beta$ according as $\alpha - \beta < 0$ or $\alpha - \beta > 0$. It becomes clear that with these conventions, Rules 1-16 in Section 13 are valid. However, for this set T Archimedes' Rule 17 in Section 13 is not valid, because regardless of how large the positive number A may be chosen there always remains $At < 1$. This number set T is non-Archimedean.

If τ is any expression of the form

$$\tau = a_0 t^n + a_1 t^{n+1} + a_2 t^{n+2} + \ldots,$$

where a_0 ($\neq 0$), a_1, $a_2, \ldots$ are arbitrary real numbers and n is the lowest positive exponent of t, then τ will be called *an infinitesimally* small number of the set T.

Any power series of the form

$$\varphi(\tau) = c_0 + c_1 \tau + c_2 \tau^2 + \ldots,$$

in which c_0, c_1, $c_2, \ldots$ are any real numbers and τ is an infinitesimally small number of the set T is again a number of the set T. Indeed it can be ordered according to increasing powers of the parameter t whereby every coefficient can be obtained as a real number in a finite number of operations.

Furthermore, if α and β are any two numbers of the set T then

$$\alpha + i\beta$$

will be called *an element of a set of complex numbers T* where i is the imaginary unit, i.e., $i^2 = -1$ and $\alpha + i\beta = \alpha' + i\beta'$ means $\alpha = \alpha', \beta = \beta'$.

If the functions $\sin \tau$, $\cos \tau$, e^{τ}, $e^{i\tau}$ of an infinitesimally small number τ are defined by their power series then the functional values are again numbers of the set T, or complex numbers of that set. Now if ϑ is a real number it is possible to define the functions $\sin(\vartheta + \tau)$, $\cos(\vartheta + \tau)$, $e^{i(\vartheta+\tau)}$, $e^{i\vartheta+(1+i)\tau}$ in the set T by the formulas

$$\begin{aligned}
\sin(\vartheta + \tau) &= \sin\vartheta \cos\tau + \cos\vartheta \sin\tau, \\
\cos(\vartheta + \tau) &= \cos\vartheta \cos\tau - \sin\vartheta \sin\tau, \\
e^{i(\vartheta+\tau)} &= e^{i\vartheta} e^{i\tau} \\
e^{i\vartheta+(1+i)\tau} &= e^{\tau} e^{i(\vartheta+\tau)}.
\end{aligned}$$

From these definitions one obtains the well-known relations

$$\begin{aligned}
\cos^2(\vartheta + \tau) + \sin^2(\vartheta + \tau) &= 1 \\
\cos(\vartheta + \tau) \pm i \sin(\vartheta + \tau) &= e^{\pm i(\vartheta+\tau)}.
\end{aligned}$$

Now a geometry will be constructed with the aid of the number set T as follows:

Consider a pair of numbers (x, y) of the set T as a point, and the ratio $(u : v : w)$ of any three numbers from T as a line if u and v are not both 0. Furthermore, let the existence of the equation

$$ux + vy + w = 0$$

express the fact that the point (x, y) lies on the line $(u : v : w)$.

A plane geometry that is constructed in the given way from a number complex in which Rules 1-16 in Section 13 hold, always satisfies Axioms I, 1-3 and IV, as has already been mentioned in Section 9.

It is easy to see that a line is also determined by one of its points (x_0, y_0) and the ratio of two numbers α, β both of which do not vanish. The equation

$$x + iy = x_0 + iy_0 + (\alpha + i\beta)s; \quad (\alpha + i\beta \neq 0),$$

in which s denotes any number from the set T characterizes the incidence of the point (x, y) on the given line. Let the points of a line be ordered according to the magnitude of the parameter s. Then a ray emanating from the point (x_0, y_0) on the given line is determined through the auxiliary condition $s > 0$ or $s < 0$. If to two points A and B on a line there correspond the parametric values s_a and s_b $(> s_a)$

then the segment AB is represented by the equation of the line and the auxiliary condition $s_a \leqq s \leqq s_b$. Axioms II, 1-3 are satisfied now. In order to see that the axiom of order II, 4 is also satisfied let the following convention be made: Let a point (x_3, y_3), lie on one side or on the other side of the line determined by the points (x_1, y_1) and (x_2, y_2) according as the sign of the determinant

$$\begin{vmatrix} x_2 - x_1 & y_2 - y_1 \\ x_3 - x_1 & y_3 - y_1 \end{vmatrix}$$

is positive or negative, respectively. It is easy to convince oneself that the definition of the side with respect to a line given in this way does not depend on the choice of the points (x_1, y_1) and (x_2, y_2) on the line and agrees with the definition of the side given on p. 8.

For the definition of congruence take the transformation

$$x' + iy' = e^{i\vartheta + (1+i)\tau} (x + iy) + \lambda + i\mu,$$

which can be written briefly in the form

$$x' + iy' = [\vartheta, \tau; \lambda + i\mu] \; (x + iy)$$

where ϑ is any real number, τ an infinitesimally small number of the set T and λ, μ denote any two numbers from the set T. A transformation of this form will be called a *congruent mapping*. A congruent mapping with vanishing λ, μ will be called a rotation about the point (0, 0).

The totality of these congruent mappings forms a **group**, i.e., it has the following four properties:

1. **There exists a congruent mapping which leaves all points fixed:**

$$[0, 0; 0] \; (x + iy) = x + iy.$$

2. **If two congruent mappings are performed consecutively then the result represents a congruent mapping:**

$$[\vartheta_2, \tau_2; \lambda_2 + i\mu_2] \; \{[\vartheta_1, \tau_1; \lambda_1 + i\mu_1] \; (x + iy)\}$$
$$= [\vartheta_2 + \vartheta_1, \tau_2 + \tau_1; \lambda_2 + i\mu_2 + e^{i\vartheta_2 + (1+i)\tau_2} (\lambda_1 + i\mu_1)] \; (x + iy).$$

For every congruent mapping there exists an inverse:

$$[-\vartheta, -\tau; -(\lambda + i\mu) \, e^{-i\vartheta - (1+i)\tau}] \; \{[\vartheta, \tau; \lambda + i\mu] \; (x + iy)\} = x + iy.$$

This property is a consequence of properties 1, 2, 4, 5.

The operation of congruent mapping is associative, i.e., denoting three congruent mappings by K_1, K_2, K_3 and the one resulting from

$K_1 K_2$, according to 2, by K_2K_1 then

$$K_3(K_2K_1) = (K_3K_2)K_1$$

always holds.

Besides these let the following properties of the congruent mapping be pointed out.

3. **A point is always carried again into a point of the geometry.**

The pair of numbers x', y', which arises from the pair of points x, y of T under a congruent mapping is again always in the set T.

4. **A line is orderly carried back to a line.**

It is easy to deduce the relation

$$[\vartheta, \tau; \lambda + i\mu] \left\{x_0 + iy_0 + (\alpha + i\beta)\, s\right\} = x_0' + iy_0' + (\alpha' + i\beta')\, s,$$

in which, since the exponential function never vanishes, from $\alpha + i\beta \neq 0$, it always follows that $\alpha' + i\beta' \neq 0$.

As an immediate result one obtains the following: Two distinct points are always carried again into two distinct points.

5. **There exists exactly one congruent mapping which carries a given ray h into a given ray h'.**

Let h be given by the equation

$$x + iy = x_0 + iy_0 + (\alpha + i\beta)\, s, \qquad \alpha + i\beta \neq 0,\ s > 0$$

and h' by the equation

$$x' + iy' = x_0' + iy_0' + (\alpha' + i\beta')\, s', \qquad \alpha' + i\beta' \neq 0, \qquad s' > 0.$$

A congruent mapping $[\vartheta, \tau; \lambda + i\mu]$, which carries h into h' must, to begin with, carry the point from which h emanates into the point from which h' emanates,

(1) $$x_0' + iy_0' = e^{i\vartheta + (1 + i)\tau}\,(x_0 + iy_0) + \lambda + i\mu.$$

Furthermore, to every positive value of s there must correspond a positive value of s' such that

$$x_0' + iy_0' + (\alpha' + i\beta')\, s' = [\vartheta, \tau; \lambda + i\mu] \left\{x_0 + iy_0 + (\alpha + i\beta)\, s\right\}$$

and hence

(2) $$(\alpha' + i\beta')\, s' = e^{i\vartheta + (1 + i)\tau}\,(\alpha + i\beta)\, s.$$

Conversely, every congruent mapping which satisfies equations (1) and (2) carries h into h'.

Divide the last equation by its complex conjugate so that

(3) $$\frac{\alpha' + i\beta'}{\alpha' - i\beta'} = e^{2i(\vartheta + \tau)}\, \frac{\alpha + i\beta}{\alpha - i\beta}\ .$$

Setting $$\frac{\alpha' + i\beta'}{\alpha' - i\beta'} \cdot \frac{\alpha - i\beta}{\alpha + i\beta} = \xi + i\eta,$$

one obtains $$(\xi + i\eta)\ (\xi - i\eta) = \xi^2 + \eta^2 = 1.$$

ξ and η as numbers from T are power series in the parameter t. By comparing coefficients it can be concluded from the last equation that no negative powers of the parameter t can occur in them, and moreover, that they can be represented in the form

$$\xi = a + \xi', \quad \eta = b + \eta'$$

where a, b are ordinary real numbers and ξ', η' denote infinitesimally small numbers from T, and that the relations

$$(4) \quad \begin{aligned} a^2 + b^2 &= 1, \\ 2(a\xi' + b\eta') + \xi'^2 + \eta'^2 &= 0 \end{aligned}$$

hold

Equation (3) $e^{2i(\vartheta + \tau)} = \xi + i\eta$

can, with the aid of the given definition of the trigonometric functions, be cast in the form:

$$(5) \quad \begin{aligned} \cos 2(\vartheta + \tau) &= \cos 2\vartheta \cos 2\tau - \sin 2\vartheta \sin 2\tau = \xi \\ &= a + \xi', \\ \sin 2(\vartheta + \tau) &= \sin 2\vartheta \cos 2\tau + \cos 2\vartheta \sin 2\tau = \eta \\ &= b + \eta'. \end{aligned}$$

By comparing coefficients these equations lead to the equations

$$\cos 2\vartheta = a, \quad \sin 2\vartheta = b,$$

from which, on the basis of the validity of the equation $a^2 + b^2 = 1$, the real number ϑ can be uniquely determined to within an integral multiple of π. Substituting the values $\cos 2\vartheta = a$, $\sin 2\vartheta = b$ in equation (5) the relations

$$\cos 2\tau = 1 + a\xi' + b\eta', \quad \sin 2\tau = a\eta' - b\xi'$$

can be deduced. Since by equations (4) the sum of the squares of the right-hand sides is 1 the infinitesimally small number τ is therefore uniquely determined. It can be computed by comparing coefficients in one of the last two equations.

Since ϑ is determined to within an integral multiple of π the factor $e^{i\vartheta + (1+i)\tau}$ is only determined to within a sign. As can be easily seen only one of the two signs yields in equation (2) a positive s' for a positive s. Hence the real number ϑ is determined to within an integral multiple of 2π. Substitution of the pair of values ϑ, τ in equation (1) determines uniquely also the numbers λ and μ in T. Finally upon some reflection one concludes that equations (1) and (3), and hence the computed values ϑ, τ, λ, μ, are independent of the type of representation of the rays h and h'.

6. **For every two points A and B there always exists a congruent mapping that carries A into B and B into A.**

If the points A, B have the coordinates x_1, y_1 and x_2, y_2, respectively, then the congruent mapping

$$[\pi, 0;\ x_1 + x_2 + i(y_1 + y_2)]$$

yields the desired transformation.

7. **If a congruent mapping carries a ray h into a ray h' and a point P on the right or the left of h into a point P' then P' also lies on the right or the left of h', respectively. In brief, P and P' are equipositional with respect to h and h', respectively.**

It will be shown next that the determinants

$$\begin{vmatrix} x_2 - x_1 & y_2 - y_1 \\ x_3 - x_1 & y_3 - y_1 \end{vmatrix}, \quad \begin{vmatrix} x'_2 - x'_1 & y'_2 - y'_1 \\ x'_3 - x'_1 & y'_3 - y'_1 \end{vmatrix}$$

have the same sign when and only when the points (x_3, y_3) and (x'_3, y'_3) are equipositional with respect to the oriented lines that are determined by the points $(x_1, y_1), (x_2, y_2)$ and $(x'_1, y'_1), (x'_2, y'_2)$, respectively (cf. p. 113). Next one deduces from the definition of "right" and "left" given on p. 64 that the points (x_3, y_3) and (x_2, y_2) are not equipositional with respect to the oriented lines which are determined by the points (x_1, y_1), (x_2, y_2) and (x_1, y_1), (x_3, y_3), respectively. In fact, the corresponding determinants differ only in sign. The asserted result follows from the fact that the definition of a side of a line by means of the sign of the given determinant satisfies the properties of a side of a line given on page 8.

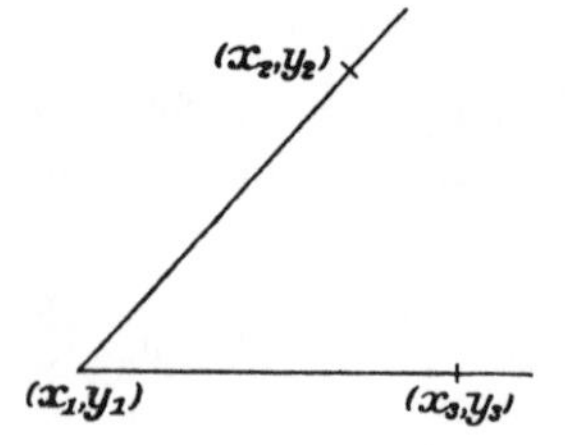

Property 7 will be demonstrated once it is shown that the sign of the determinant

$$\begin{vmatrix} x_2 - x_1 & y_2 - y_1 \\ x_3 - x_1 & y_3 - y_1 \end{vmatrix}$$

is preserved under a congruent mapping. Since the determinant differs from the imaginary part of the quotient

$$\frac{(x_3 + iy_3) - (x_1 + iy_1)}{(x_2 + iy_2) - (x_1 + iy_1)},$$

only by a positive factor it is immediately apparent that this quotient is invariant under a congruent mapping.

Let it be agreed now that a segment shall be said to be congruent to another segment if and only if there exists a congruent mapping which carries the first into the second, and that an angle shall be said to be congruent to another angle if and only if there exists a mapping which carries the first into the second.

The following will be shown now:

The above definition of congruence of segments and angles satisfies Axioms III, 1-6 as soon as the assumed congruent mapping has properties 1 through 7.

The validity of Axiom III, 1 is an immediate consequence of property 5.

The validity of Axiom III, 2 can be shown in the following manner: Let the congruent mappings K_1 and K_2 carry the segment $A'B'$ and $A''B''$ into the segment AB, respectively. From properties 1, 2, 4, 5 it follows that for the congruent mapping K_2 there exists an inverse congruent mapping K_2^{-1}. The mapping $K_2^{-1}K_1$ which exists by property 2 carries the segment $A'B'$ into $A''B''$.

The validity of Axiom III, 6 can be proved in a similar way.

It will be shown now that if a segment AB is congruent to a segment $A'B'$ then the congruent mapping K which carries the ray AB into the ray $A'B'$ also carries B into B'. Let the congruence of the segments AB and $A'B'$ be established by a congruent mapping K_1. In case K_1 carries the point A into A' then according to property 4 the congruent mapping KK_1^{-1} carries the ray $A'B'$ into itself and so by properties 1 and 5 must be the identity. However, in case K_1 carries A into B' take the congruent mapping K_2 that carries A into B and B into A, which exists according to property 6. The congruent mapping $K(K_2K_1^{-1})$ carries now the ray $A'B'$ into itself and so is the identity.

From the results shown here and from properties 4 and 5 the validity of Axiom III, 3 follows immediately and from these results and properties 4, 5 and 7 the validity of Axiom III, 5* also follows immediately.

Finally, the validity of Axiom III, 4 can be shown in the following manner: If an angle $\measuredangle (a,b)$ and a ray c are given then by property 5 there exists exactly one congruent mapping K_1 which carries a into c and exactly one congruent mapping K_2 which carries b into c. K_1 carries b into a ray b' that is distinct from c, as can be seen from property 4 by considering the congruent mapping K_1^{-1}. In like manner K_2 carries the ray a into the ray a' that is distinct from c. The congruent mapping $K_2K_1^{-1}$ carries c into a' and b' into c. It follows now from property 7 that a' and b' lie on different sides of c. The first part of Axiom III, 4 is thus satisfied. The second part is an immediate consequence of property 1.

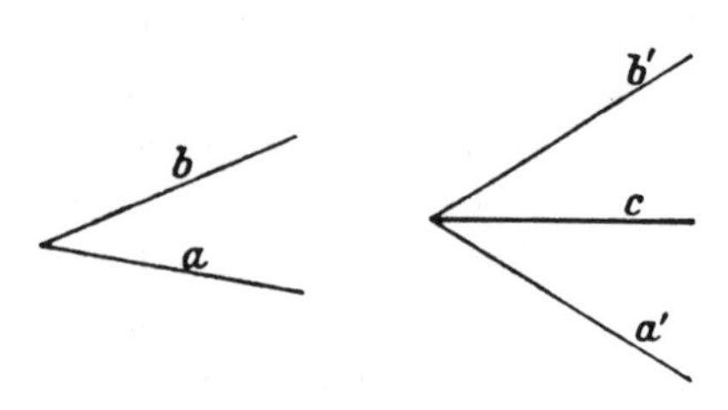

The validity of Axiom III, 7 becomes clearer from the following considerations: A ray that emanates from the point (0, 0), which will be denoted by O, can be represented by an equation of the form

$$x + iy = e^{i(\vartheta + \tau)} s; \; s > 0$$

and is obtained from the positive x-axis by the rotation $[\vartheta, \tau, 0]$. Of the two rays that emanate from O and extend into the upper half plane the one which has the smaller sum $\vartheta + \tau$ modulo 2π, as is easy to see, lies between the other and the positive x-axis.

Now let the right side h of an angle coincide with the positive x-axis. Let the left side k be represented by the equation

$$x + iy = e^{i(\vartheta_1 + \tau_1)} s; \; s > 0.$$

Introduce into the interior of this angle a ray h' that emanates from O. Then there exists exactly one congruent mapping which carries h into h' namely, a rotation $[\vartheta_2, \tau_2; 0]$. It carries k into the ray k' by the equation

$$x + iy = e^{i(\vartheta_1 + \vartheta_2 + \tau_1 + \tau_2)} s; \; s > 0.$$

Since

$$\vartheta_1 + \vartheta_2 + \tau_1 + \tau_2 > \vartheta_1 + \tau_1 \qquad \text{mod } 2\pi;$$

k' does not lie inside angle $\measuredangle(h, k)$.

The validity of the neighborhood Axiom V, 3 can be proved in the following manner: It is easy to show with the aid of the second congruence theorem and Axiom IV that for every segment lying in the interior of a triangle it is possible to find a congruent segment which, emanating from a vertex, either lies on a side or in the interior of the triangle.

By Axiom III, 1 there exists for a given segment AB exactly one segment OB' which emanates from the point O in the direction of the positive x-axis to which the segment AB is congruent. Let the abscissa β of B' denote the *length* of the segment AB,

$$\overline{AB} = \beta .$$

Consider now the triangle with the vertices $O\,(0,0)$, $C\left(\frac{\beta}{2}, 0\right)$, $D\left(\frac{\beta}{4}, \frac{\beta}{4}\sqrt{3}\right)$. This triangle is equilateral and its angles are equal, as is shown by the congruent mapping $\left[\frac{2\pi}{3}, 0; \frac{\beta}{2}\right]$, which carries O into C, C into D and D into O. The free endpoint F of a segment which, emanating from O either lies on a side or in the interior of the angle $\measuredangle COD$ and which is congruent to AB, can be represented in the form

$$[\vartheta, \tau; 0]\beta, \quad 0 \leqq \vartheta + \tau \leqq \frac{\pi}{3}.$$

However, all points which can be represented in this form lie on the side of the line CD on which O does not lie. This can be seen by substituting the coordinates of O and F into the determinant

$$\begin{vmatrix} 1 & -\sqrt{3} \\ x_3 - \frac{\beta}{2} & y_3 \end{vmatrix}$$

which according to p. 116 corresponds to CD. It has thus been shown that no segment congruent to AB can be found in the interior of the triangle OCD.

This is summarized as follows:

In the given geometry all axioms of ordinary geometry formulated above with the exception of the Archimedean Axiom V, 1 are valid. In this assertion the congruence axiom is to be understood in the narrower form III, 5*.

Furthermore, the following theorem holds:

Every angle can be bisected and there exists a right angle.

It is sufficient to show that every angle originating at O can be bisected. Let $[\vartheta, 0; 0]$ be the rotation which carries the right side into the left side of the angle. The rotation $\left[\frac{\vartheta}{2}, \frac{\tau}{2}; 0\right]$ carries the right side into the bisector side.

The existence of a right angle can be deduced by considering the rotation $\left[\frac{\pi}{2}, 0; 0\right]$

The concept of *reflection* in a line a will be introduced now as follows: If a perpendicular to a line a is dropped from any point A and extended an equal amount to itself about its foot B to A' then A' is called the image point of A. Reflect now a point A with coordinates, $\alpha > 0, \beta > 0$ in the positive x-axis. Let the angle $\measuredangle AOB$ between the ray OA and the positive x-axis be $\vartheta + \tau$ and moreover, let the point $x = \gamma$ on the x-axis be carried into the point A by a rotation through the angle $\vartheta + \tau$ so that

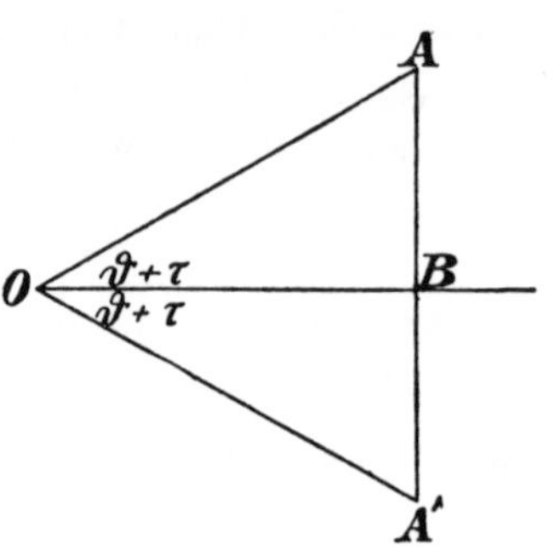

$$e^{i\vartheta + (1+i)\tau}\gamma = \alpha + i\beta.$$

The coordinates of the image A' of the point A with respect to the x-axis are $\alpha, -\beta$. If the rotation about the angle $\vartheta + \tau$ is performed there results from the point A' a point which is represented by the complex number

$$e^{i\vartheta + (1+i)\tau}(\alpha - i\beta) = \frac{\alpha + i\beta}{\gamma}(\alpha - i\beta) = \frac{\alpha^2 + \beta^2}{\gamma}$$

i.e., the desired point lies on the positive x-axis. Consequently the angle $\measuredangle A'OB$ is equal to $\vartheta + \tau$ and thus is also equal to the angle $\measuredangle AOB$. The result is expressed as follows:

If in two symmetrically situated right triangles two legs are equal then the corresponding angles on the hypotenuses are equal.

The following general theorem is deduced at the same time:

In the image of a figure the angles are equal to the corresponding angles of the original figure.

The fundamental theorem of the theory of proportion (Theorem 42) as well as Pascal's Theorem (Theorem 40) can be deduced with no difficulty from the fact that in the given geometry lines are defined by linear equations. From this the following result can be seen:

In the given geometry the theory of proportion is valid and furthermore, all theorems of affine geometry hold in it (cf. Section 35).

By the validity of Axiom III, 7 it can be shown that in the given geometry angles can be compared by their magnitudes in a uniquely determined way.

With the aid of this result the **exterior angle theorem** (Theorem 22) can be proved, and indeed since in the given geometry vertical angles are always equal, the proof given on pages 21-22 can be carried over here. From the result that in the given geometry the sum of two angles can be defined uniquely, one obtains with the aid of Axiom IV the **angle sum theorem for a triangle** (Theorem 31).

One approaches now the fundamental question whether in the given geometry the theorem on the equality of the base angles in an isosceles triangle holds (Theorem 11).

From this and the exterior angle theorems one obtains by an indirect proof the converse of the base angle theorem (Theorem 24) and, with the aid of a well-known theorem of Euclid, the following theorem: The sum of two sides in a triangle is greater than the third. However neither of these two theorems, as will be shown, holds in the given geometry; thereupon, it will be shown at the same time that the base angle theorem does not hold in it.

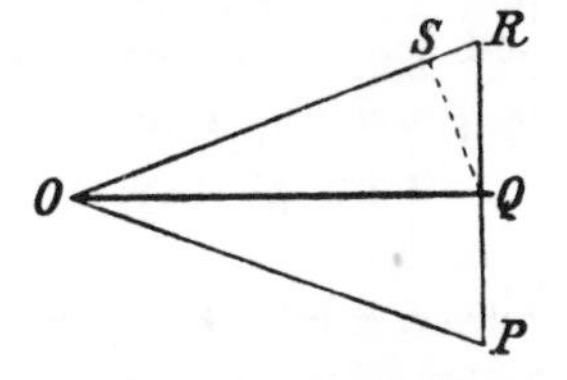

Consider the triangle OQP whose vertices have the coordinates $(0, 0)$, $(\cos t, 0)$, $(\cos t, -\sin t)$. The lengths (see p. 123) of the segments OP and QP are obtained by the congruent mappings $[0, t; 0]$ and $\left[\frac{\pi}{2}, 0; -\cos t \cdot e^{i\frac{\pi}{2}}\right]$ respectively. One obtains

$$\overline{OP} = e^t = 1 + t + \frac{t^2}{2} + \cdots,$$

$$\overline{QP} = \sin t = t - \frac{t^3}{6} + - \cdots,$$

$$\overline{OQ} = \cos t = 1 - \frac{t^2}{2} + - \cdots.$$

It can be seen from the definition of order for the numbers of the set T that

$$\overline{OQ} + \overline{QP} < \overline{OP}.$$

The theorem according to which the sum of two sides in a triangle is greater than the third does not hold in the given geometry.

From this result the fundamental dependence of this theorem on the triangle congruence axiom in the broader sense can be realized.

By this result the following holds at the same time:

In the given geometry the isosceles triangle theorem does not hold and hence neither does the triangle congruence axiom in the broader sense.

That the converse of the base angle theorem does not hold either can be seen from the example of the triangle OPR where R is the reflection image of the point P with respect to the line OQ, i.e., its coordinates are ($\cos t$, $\sin t$). Then by a theorem proved above (p. 124)

$$\measuredangle\ OPR \equiv \measuredangle\ ORP.$$

Neverthless the sides OP and OR are not congruent to each other. The length of the segment OR, which is computed through the rotation $[0, -t; 0]$ is

$$\overline{OR} = e^{-t} \neq \overline{OP} = e^{t}.$$

It can be seen from this **that in general the hypotenuses of two symmetrically situated right triangles with coinciding legs are different and hence the images of segments under a reflection in a line are not necessarily equal to those in the original figure.**

In the given geometry, as W. Rosemann[1] has shown, **the third congruence theorem** (Theorem 18) **does not even hold in the narrower form restricted to equiposition triangles.**

In order to realize this, note that the points $A = 0$, $B = t$, $C = te^{i\frac{\pi}{3}}$ determine an equilateral triangle. Examing next the point

$$D = \frac{t}{1 - e^{(1+i)t}},$$

it can be seen that $AD \equiv BD$ as the congruent mapping $[0, t; t]$ carries D into itself and A into B. It can be established that the points A and B lie on the same side of the line CD. From this it first follows that the triangles ACD and BCD, which are equal in all their sides, are equipositioned triangles, and second that they cannot be equal in all their angles.

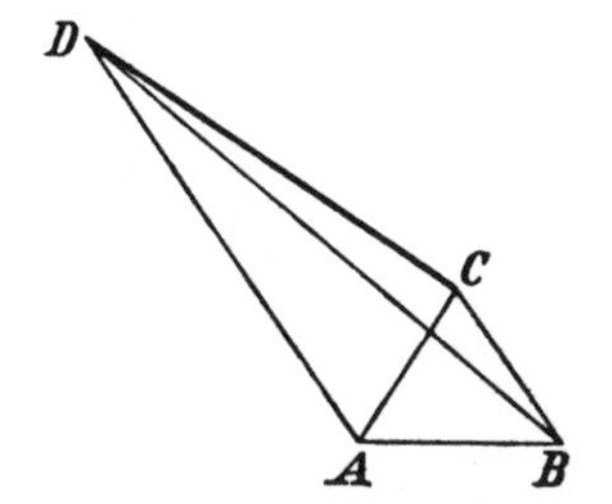

The Euclidean theory of area of polygons in the given geometry will be discussed next. This theory was developed in Section 20 on the concept of the *area* of a triangle. The proof that this area, one half the product of the base and the altitude, is independent of whichever side of the triangle is considered as the base, will rest on an application of the triangle congruence axiom to symmetrically situated triangles. That

[1] "Der Aufbau der ebenen Geometrie ohne das Symmetrieaxiom," Dissertation, *Math. Ann.*, Vol. 90 (Göttingen, 1922). There for the first time is also shown the dependence of the validity of Axioms III, 1-6 on certain properties of the congruent mapping.

it cannot be proved without the broader form of this axiom can be seen from the example of the triangle OQR on page 125. QR is the altitude to OQ. The length

$$\overline{QR} = \sin t$$

is obtained by the congruent mapping $\left[-\frac{\pi}{2}, 0; -\cos t \cdot e^{-i\frac{\pi}{2}}\right]$ Since $\overline{OQ} = \cos t$ the area, on one hand, must be

$$J = \frac{\cos t \cdot \sin t}{2}.$$

On the other hand the foot S of the perpendicular or OR dropped from Q is computed to be

$$S = \cos t + ie^{it} \sin t \cos t.$$

Now the length

$$\overline{QS} = e^{-t} \sin t \cos t$$

is obtained by the congruent mapping

$$\left[-\frac{\pi}{2}, -t; -\cos t \cdot e^{-i\frac{\pi}{2}-(1+i)t}\right].$$

Since $\overline{OR} = e^{-t}$ one obtains

$$J = \frac{e^{-t} \cdot e^{-t} \cos t \sin t}{2}$$

for the area, which is certainly less than

$$\frac{\cos t \sin t}{2}.$$

Whereas the concept of area loses its meaning without the broader form (III, 5) of the axiom of triangle congruence, the concepts of *equidecomposability* and *equicomplementability* of polygons can be defined exactly as in Section 18. One then obtains Theorem 46 exactly as in Section 19 which expresses the **equicomplementability of two triangles of equal bases and altitudes.**

Furthermore it can be seen that by the narrower form of Axiom III, 5*[1] also a square, i.e., a rectangle of equal sides, can be constructed on every segment. **Pythagoras' Theorem, according to which the squares on the two legs of any right triangle together are equicomplementable with the square of the hypotenuse, holds now in the given geometry.** This follows because throughout the Euclidean proof of Pythagoras' Theorem, only the congruence of equipositioned

[1] The axiom of parallels, and the existence of a right angle, are also required.

triangles, and hence only the triangle congruence axiom in the narrower sense, was used.[1]

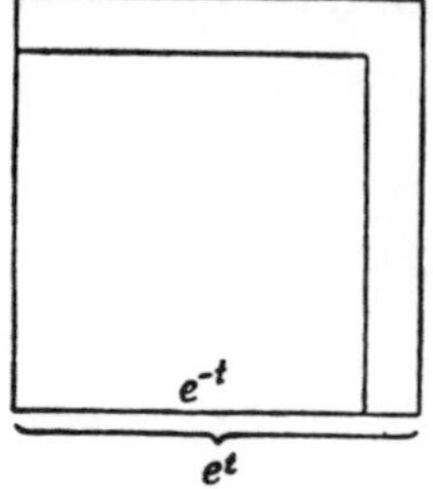

An application of Pythagoras' Theorem to the triangles OQP and OQR on page 125 shows, with the aid of Theorem 43, that the squares constructed on the segments OP and OR **are equicomplementable although the segments themselves**, as computed above, **are not equal.**

The relation of this situation to Theorem 52 is clear and the following can be seen from it: **Euclid's fundamental theorem according to which two equicomplementable triangles with equal bases have equal altitudes also has no validity in the given geometry.**

In fact, Theorem 48 was proved in Sections 20 and 21 by essentially using the concept of area.

The given geometry leads to the following result:

It is impossible to base the Euclidean theory of area on the triangle congruence axiom in the narrower sense even by assuming the validity of the theory of proportion.

Since in the given geometry the well-known relation between the hypotenuse and the legs of a right triangle, which in ordinary geometry is deduced from Pythagoras' Theorem, does not hold, this geometry is called here a *non-Pythagorean geometry.*

The main results of this non-Pythagorean geometry are summarized as follows:

If the triangle congruence axiom is taken in the narrower sense and if of the continuity axioms only the neighborhood axiom is taken as valid then it is impossible to prove the base angle theorem for isosceles triangles, even if the validity of the theory of proportion is assumed. Nor does the Euclidean theory of area follow. Even the theorem according to which the sum of two sides of a triangle is greater than the third and the third congruence axiom for equipositioned triangles are not necessary consequences of the assumption made.

Another non-Pythagorean geometry will be constructed which differs from the one treated above by the fact that in it the Archimeadean Axiom V, 1 holds but not the neighborhood axiom V, 3.

This geometry will be based on the sub-field Ω of the real numbers which is generated from the numbers 1 and $\pi = \tan 1$ by a finite

[1] See also Suppl. V, 1.

number of applications of the operations of addition $\omega_1 + \omega_2$, subtraction $\omega_1 - \omega_2$, multiplication $\omega_1 \cdot \omega_2$, division $\omega_1 \div \omega_2$ (in case $\omega_2 \neq 0$) and exponentiation $\omega_1^{\omega_2}$.[1] Here ω_1, ω_2 denote numbers which have already been obtained by the five operations from the numbers 1 and τ. In order to obtain the number ω from 1 and τ let the first operation be applied n_1 times, the second n_2 times . . . . the fifth n_5 times. The numbers ω of the field Ω can then be enumerated according to the increasing sums $n_1 + n_2 + \ldots + n_5$.

Let a plane geometry be constructed on this number set by the same conventions that the non-Pythagorean geometry on page 116 was constructed on the number set T. Just as there the validity of Axioms I, 1-3, II, IV in this geometry can be seen from the fact that defining order in a natural way all Rules of Operation 1-16 of Section 13 hold in Ω.

For every number ω of the field Ω extended by the number ∞ there exist infinitely many numbers ϑ which satisfy the equation

$$\vartheta = \arctan \omega$$

The totality of numbers ϑ obtained from Ω through this equation form a field Θ which does not coincide with Ω, but which is countable as Ω. Take any countable subset of Θ. In it there exists a first number that is not a multiple of π. Denote it by ϑ_{k_1}. Let the first number in Θ which cannot be represented in the form $\vartheta = r\pi + r_1 \vartheta_{k_1}$, where r, r_1 are any rational numbers, if it exists at all, be denoted by ϑ_{k_2}. Continuing in this manner let the first number ϑ in Θ which cannot be represented in the form

$$\vartheta = r\pi + r_1\vartheta_{k_1} + r_2\vartheta_{k_2} + \ldots + r_n\vartheta_{k_n},$$

if such a number exists at all, be denoted by $\vartheta_{k_{n+1}}$. A sequence ϑ_{k_1}, ϑ_{k_2}, ϑ_{k_3}, . . . is thus defined which certainly contains one term and perhaps an infinite number of terms. Every number ϑ in Θ can be uniquely represented in the form

$$\vartheta = r\pi + r_1\vartheta_{k_1} + r_2\vartheta_{k_2} + \ldots + r_n\vartheta_{k_n},$$

where $\vartheta_{k_1}, \vartheta_{k_2}, \ldots, \vartheta_{k_n}$ are the first n terms of the sequence defined above and $r, r_1, r_2, \ldots, r_n$ are some rational numbers.

[1] Exponentiation is for positive ω_1 only. Instead $\omega_1^{\omega_2}$, $\omega_1^{\frac{1}{k}}$ (k a natural number) can do as well.

Let segment and angle congruence be defined exactly as in the first non-Pythagorean geometry on page 121 by a **congruent mapping**. Let any transformation of the form

$$x' + iy' = 2^{r_1} e^{i\vartheta} (x + iy) + \lambda + i\mu,$$

be called here a congruent mapping where ϑ is a number from Θ, r_1 is a rational number that occurs in the representation of ϑ above and λ, μ are any numbers from Ω.

As it is easy to see, the congruent mappings form a group, and thus have properties 1 and 2 introduced on page 117. Property 3 can be obtained from the fact that the numbers

$$2^{r_1}, \quad \cos\vartheta = \frac{1}{\sqrt{1+\tan^2\vartheta}}, \quad \sin\vartheta = \frac{\tan\vartheta}{\sqrt{1+\tan^2\vartheta}}$$

are numbers of the field Ω. Property 5 is obtained in the following way:

The proof, as on page 118, will be reduced to determining uniquely a ϑ of the field Ω to within an integral multiple of 2π, from the equation

$$2^{r_1} e^{i\vartheta} = \frac{\alpha' + i\beta'}{\alpha + i\beta} \cdot \frac{s'}{s}.$$

Dividing the imaginary part by the real part

$$\tan\vartheta = \frac{\alpha\beta' - \beta\alpha'}{\alpha\alpha' + \beta\beta'}.$$

The number ϑ is determined in the number set Θ by this equation, to within an integral multiple of π. As in the first non-Pythagorean geometry the determination is to within an integral multiple of 2π (cf. p. 119). The proofs of properties 4, 6 and 7 proceed exactly as there.

From the seven properties of congruent mappings thus proved it follows by the general proof given on page 121 that Axioms III, 1-6 are satisfied in this geometry. The validity of Axiom III, 7 follows in an obvious way similar to the one in the first non-Pythagorean geometry.

With the aid of the definitions of order and congruence, the validity of the Archimedean Axiom V, 1 follows from the fact that the field Ω is a subfield of the field of real numbers.

That the neighborhood axiom, however, does not hold can be shown in the following way: For every triangle it is possible to find a congruent triangle OAB whose vertices are $O = (0, 0)$, $A = (\alpha, 0)$, $B = (\beta, \gamma)$ where α and γ denote positive numbers. It is therefore sufficient to show that a segment of, say, unit length lies in every such triangle. Whether β vanishes or not the ray OB can be represented in the form

$$x + iy = e^{i \arctan \frac{\gamma}{\beta}} \cdot s,$$

where s denotes a positive parameter from Ω. Since $\alpha\gamma$ and $|\alpha - \beta| + \gamma$ are positive it is possible to find an integer r_1, that is not necessarily positive, which satisfies the inequality

$$(1) \qquad 2^{r_1} < \frac{\alpha\gamma}{|\beta - \alpha| + \gamma}.$$

Now for the given numbers r_1, ϑ_{k_1}, $\arctan\frac{\gamma}{\beta} > 0$ there certainly exist two intergers a and b which satisfy the inequality

$$(2) \qquad 0 < \frac{a}{2^b}\pi + r_1\vartheta_{k_1} < \arctan\frac{\gamma}{\beta}.$$

From the formula

$$\tan\frac{\vartheta}{2} = \frac{-1 \pm \sqrt{1 + \tan^2\vartheta}}{\tan\vartheta}$$

it is seen that $\frac{\pi}{2^b}$, and hence, by the addition formula of the tangent function, that also

$$\vartheta = a\frac{\pi}{2^b} + r_1\vartheta_{k_1}$$

are numbers in the field Θ. From inequality (2) it follows that the ray

$$x + iy = e^{i\vartheta} \cdot s, \quad s > 0$$

lies in the interior of the angle $\measuredangle AOB$. The free point C of the unit segment emanating from O and lying on this ray can be represented in the form

$$x + iy = 2^{r_1} \cdot e^{i\vartheta}.$$

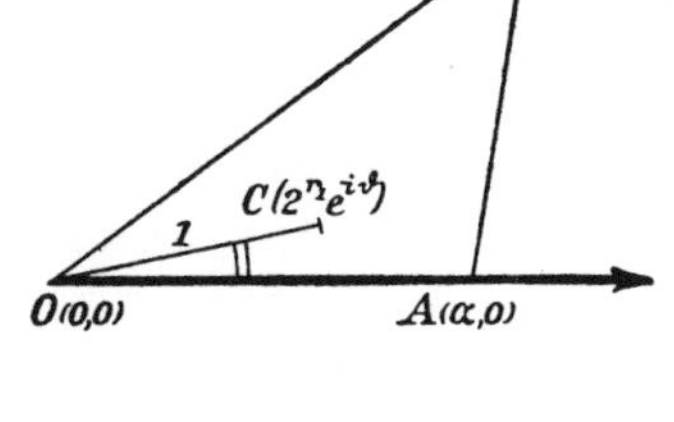

The points C and O lie on the same side of the line AB since both determinants

$$\begin{vmatrix} \beta - \alpha & \gamma \\ -\alpha & 0 \end{vmatrix} = \alpha\gamma,$$

$$\begin{vmatrix} \beta - \alpha & \gamma \\ 2^{r_1}\cos\vartheta - \alpha & 2^{r_1}\sin\vartheta \end{vmatrix} > -2^{r_1}|\beta - \alpha| - 2^{r_1}\gamma + \alpha\gamma$$

are positive, the last one by inequality (1). Hence C lies in the interior of the triangle OAB; i.e., there exists in the interior of this triangle a segment of unit length.

The possiblity of bisecting an angle and the existence of a right angle can be shown exactly as in the first non-Pythagorean geometry. The theorems on the images under reflection as well as all theorems of the theory of proportion and of affine geometry introduced on page 124 and page 126 can be shown to be valid in a similar way. All angles of this geometry also occur in Euclidean geometry and the ordering by magnitude is the same as there. From this also follow the validity of the exterior angle theorem (Theorem 22) and the theorem about the sum of the angles in a triangle. However, the equal base angle theorem in an isosceles triangle does not hold. From this theorem, with the aid of the exterior angle theorem, as has already been noted on page 124, follows immediately its converse. That this converse does not hold in this geometry can be seen, for example, by considering the triangle *OPR* with the vertices $O = (0, 0), P = (\cos \vartheta_{k_1}, - \sin \vartheta_{k_1}), R = (\cos \vartheta_{k_1}, + \sin \vartheta_{k_1})$, which has equal angles at *P* and *R* although the lengths $\overline{OP} = 2$ and $\overline{OR} = 2^{-1}$ do not coincide.

The theory of Euclidean area is not valid either. In the same way the theorem that the sum of two sides of a triangle is greater than the third is not valid. For from this theorem it would follow immediately that every segment lying in the interior of a triangle is less than the perimeter, and thus the neighborhood axiom V, 3 would hold.

The analyzed non-Pythagorean geometries lead to the following fact:

For the proof of the validity of the equal base angle theorem in an isosceles triangle neither the Archimedean Axiom V, 1 nor the neighborhood axiom V, 3 can be dropped.

Additions to this appendix are to be found in Supplement V, 1 and 2.

APPENDIX III

A NEW DEVELOPMENT OF BOLYAI-LOBACHEVSKIAN GEOMETRY

(from *Math. Ann.*, Vol. 57)

In my festschrift, *Grundlagen der Geometrie,** Chapter I (pp. 2-32)[1] I have formulated a set of axioms for Euclidean geometry and then shown that the development of plane Euclidean geometry was possible with the plane axioms of that set alone, even if the use of the continuity axiom were avoided. In the following investigation I replace the axiom of parallels by a corresponding requirement of Bolyai-Lobachevskian geometry, and then show *that it is possible to develop Bolyai-Lobachevskian geometry in the plane exclusively with the plane axioms without the use of the continuity axioms.*[2] This new development of Bolyai-Lobachevskian geometry, as it appears to me, is not inferior, because of its simplicity, to the hitherto well-known development schemes, namely, those of Bolyai and Lobachesvsky, who both used the limiting sphere, and that of F. Klein by means of the projective method. Those developments essentially use space as well as continuity.

*This refers to the First Edition of this book which, with some revisions, was translated by E. J. Townsend as *The Foundations of Geometry* (La Salle, Ill.: The Open Court Publishing Co., 1902). (Translator's note)

[1] Compare also my article in Appendix II of this book, "Über den Satz von der Gleichheit der Basiswinkel im gleichschenklingen Dreieck," *Proceedings of the London Mathematical Society*, Vol. 35 (1903).

[2] In the meantine, the corresponding problem has also been investigated independently of Axiom IV (p. 136) which characterizes Bolyai-Lobachevskian geometry. Then M. Dehn, in his article "Über den inhalt sphärischer Dreiecke," *Math. Ann.*, Vol. 60, developed the theory of area in plane elliptic geometry without the use of the continuity axiom. Later in his article "Begründung der elliptischen Geometrie," *Math. Ann.*, Vol. 61, G. Hessenberg succeeded in giving a proof of the point of intersection theorems in plane elliptic geometry under the same assumptions. Finally, J. Hjelmslev, in the article "Neue Begründung der ebenen Geometrie," *Math. Ann.*, Vol. 64, has shown that plane geometry can be developed without the continuity axioms, and even without any assumption about intersecting or nonintersecting lines.

For the sake of easier conprehension I list the following axioms of plane geometry used subsequently in accordance with my festschrift *Grundlagen der Geometrie,* as follows:[1]

I. Axioms of Incidence

I, 1. *For two points A, B there always exists a line a on which each of the two points A, B is incident.*

I, 2. *For two points A, B there exists* **no more than** *one line on which each of the two points A, B is incident.*

I, 3. *On every line there exists at least two points. There exist at least three points which do not lie on a line.*

II. Axioms of Order

II, 1. *If a point B lies between the points A and C then A, B, C, are three distinct points of a line and B then also lies between C and A.*

II, 2. *For two points A and B there exists at least one point B on the line AC such that C lies between A and B.*

II, 3. *Among any three points of a line there exists* **no more than** *one which lies between the other two.*

DEFINITION. The points that lie between the points A and B are also called the points of the segment AB or BA.

II, 4. *Let A, B, C be three noncollinear points and a a line in the plane A, B, C which meets none of the points A, B, C. If the line a passes through a point of the segment AB then it also passes either through a point of the segment BC or a point of the segment AC.*

III. Axioms of Congruence

DEFINITION. Every line decomposes at any one of its points into two *rays.*

III, 1. *If A, B are any two points on a line a and* A' *is a point of a line* a' *then it is always possible to find a point* B' *on a ray of the line* a'

[1] The form of Axioms I-III is taken from the present edition.

determined by A' *such that the segment* AB *is congruent or equal to the segment* $A'B'$. *Symbolically*

$$AB \equiv A'B'.$$

III, 2. *If a segment* $A'B'$ *and a segment* $A''B''$ *are congruent to a segment* AB *then the segment* $A'B'$ *is also congruent to the segment* $A''B''$.

III, 3. *Let* AB *and* BC *be two segments without common points on the line* a *and furthermore let* $A'B'$ *and* $B'C'$ *be two segments without common points on the same or on another line* a'. *If* $AB \equiv A'B'$ *and* $BC \equiv B'C'$ *then* $AC \equiv A'C'$.

DEFINITION. A pair of rays h and k emanating from a point A, which together do not form a line, is called an *angle*, and is denoted either by

$$\measuredangle (h, k) \text{ or } \measuredangle (k, h).$$

The *side* of a plane with respect to a line can be defined with the aid of Axiom II. The points of a plane that lie on the same side as k with respect to h and at the same time lie on the same side as h with respect to k are called *interior* points of the angle $\measuredangle (h, k)$. They form the *angle space* of this angle.

III, 4. *Let an angle* $\measuredangle (h, k)$, *a line* a' *and a definite side of* a' *be given. Let* h' *denote a ray of the line* a' *that emanates from the point* O'. *Then there exists one and only one ray* k' *such that the angle* $\measuredangle (h, k)$ *is congruent or equal to the angle* $\measuredangle (h', k')$, *symbolically*

$$\measuredangle (h, k) \equiv \measuredangle (h', k'),$$

and such that all interior points of the angle $\measuredangle (h', k')$ *lie at the same time on the given side of* a'.

Every angle is congruent to itself, i.e.,

$$\measuredangle (h, k) \equiv \measuredangle (h, k)$$

is always true.

III, 5. *If for two triangles* ABC *and* $A'B'C'$ *the congruences*

$$AB \equiv A'B', \quad AC \equiv A'C' \quad \text{and} \quad \measuredangle BAC \equiv \measuredangle B'A'C'$$

hold then so does

$$\measuredangle ABC \equiv \measuredangle A'B'C'.$$

The triangle congruence and the isoscles triangle theorems follow easily from Axioms I-III, and the possibility of erecting or dropping a perpendicular, as well as bisecting a given segment or a given angle, can be realized at the same time. In particular, the theorem that in every triangle the sum of two sides is greater than the third follows just as in the Euclidean case.

IV. Axiom of Intersecting and Nonintersecting Lines

The axiom which in the Bolyai-Lobachevskian geometry corresponds to the axiom of parallels in Euclidean geometry is expressed as follows:

IV. *If b is any line and A a point not on the line, then there exists through A two rays a_1, a_2 which do not form one and the same line and do not intersect the line b, while every ray emanating from A that lies in the angle space formed by a_1, a_2 does intersect b.*

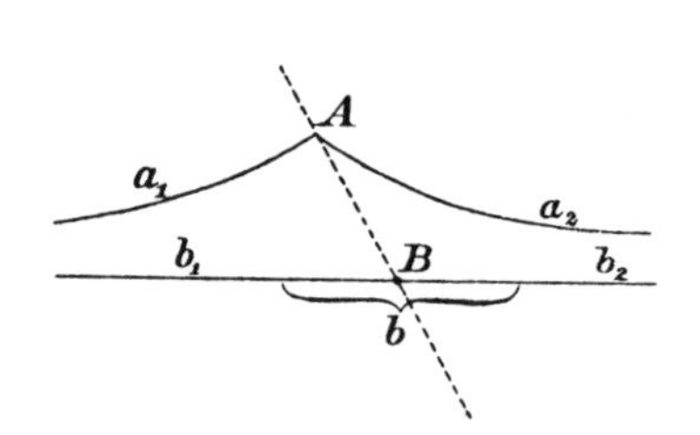

DEFINITION. Let the line b decompose into two rays b_1, b_2 at some of its points B and let a_1, b_1 lie on one side and a_2, b_2 on the other side of the line AB. Then the ray a_1 shall be said to be *parallel* to the ray b_1, and similarly the ray a_2 shall be said to be *parallel* to the ray b_2. In the same way, the rays a_1, a_2 shall be said to be *parallel* to the line b, and it will also be said that each of the two lines, of which a_1 or a_2 are rays, is *parallel* to b.

The validity of the following propositions follows immediately:

If a line or ray is parallel to another line or ray then the latter is also always parallel to the former.[1]

[1] The proof can be produced by a method due to Gauss. Cf. Bonola-Liebmann, *Die nichteuklidische Geometrie* (Leipzig, 1908 and 1921). *Non-Euclidean Geometry*, trans. by H. S. Carslaw (La Salle, Ill: The Open Court Publishing Co., 1908 and 1921).

If two rays are parallel to a third ray then they are parallel to each other.

DEFINITION. Every ray determines an end. All rays that are parallel to each other shall be said to determine the same end. Generally a ray emanating from the point A having the end α will be denoted by (A, α). A line has two ends. Generally a line whose ends are α and β will be denoted by (α, β).

If A, B and A', B' are two pairs of points and α and α' are two ends such that the segments AB and $A'B'$ are equal and moreover, the angle formed by AB and the ray (A, α) is equal to the angle formed by $A'B'$ and the ray (A', α') then, as is easy to see, the angle formed by BA and (B, α) is also equal to the angle formed by $B'A'$ and (B', α'). The two figures $AB\alpha$ and $A'B'\alpha'$ are said to be congruent.

Finally the image under reflection will be defined in the familiar way.

DEFINITION. If a perpendicular is dropped from a point to a line and is extended from its foot an equal amount to itself, then the resulting end point shall be called the *reflection image of the original point* in that line.

The reflection images of the points of a line lie again on a line. These will be called the *reflection image of the original line.*

§ 1. Lemmas

The following lemmas will be proved consecutively:

LEMMA 1. If two lines intersect a third line at equal alternate angles then they are not parallel.

PROOF. Assume to the contrary, that the two lines are parallel to each other along some direction. If half a revolution were made about the midpoint of the delineated segment on the third line, i.e., if the congruent triangle to the resulting one were constructed on the other side of that segment, it would follow that the first two lines were also parallel along the other direction, and this would contradict Axiom IV.

LEMMA 2. Given two lines a and b which neither intersect nor are parallel, there exists a line which is simultaneously perpendicular to both of them.

PROOF. From any two points A and P of the line a drop the perpendiculars AB and PB' to the line b. Let the perpendicular PB' be

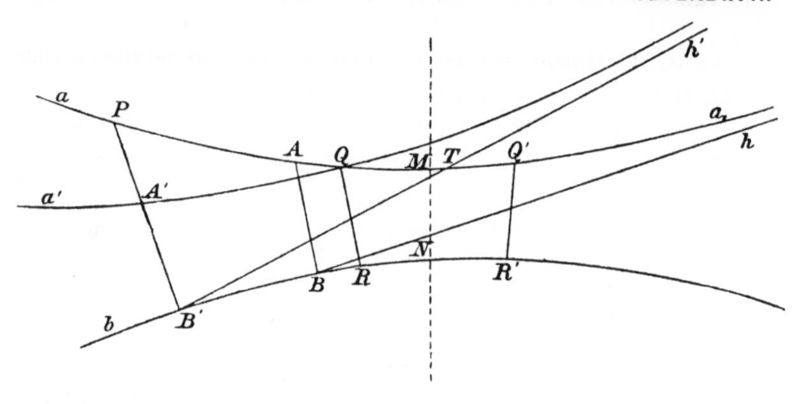

greater than the perpendicular AB. Then lay off AB on $B'P$ from B' to A' so that the point A' lies between P and B'. Now construct the line a' through A', which intersects $B'A'$ at A' at the same angle and in the same sense as the line a intersects the perpendicular BA. It will be shown that this line a' must meet the line a.

To do this denote by a_1 the ray into which a decomposes at P, and on which the point A lies, and draw from B a ray h parallel to a_1. Furthermore let h' be the ray which emanates from B' at the same angle from b and in the same direction as h. Since by Lemma 1 the ray h' is not parallel to h and hence is not parallel to a_1 and certainly does not intersect h either, then, as can readily be seen from Axiom IV, it must intersect a_1. Let T be the point of intersection of the ray h' with a_1. Since by construction A' is parallel to h' then by Axiom II, 4 the line a' must leave the triangle $PB'T$ through the side PT, and the auxiliary proof is thus complete. Let the point of intersection of the lines a and a' be denoted by Q.

From Q drop the perpendicular QR to b. Then on b lay off $B'R$ from B up to the point R' so that on b the direction from B to R' is the same as that from B' to R. In the same way lay off on a the segment $A'Q$ in the same direction from A to Q'. The connecting line MN of the midpoints M and N of the segments QQ' and RR' respectively, yields then the desired common perpendicular to a and b.

From the congruence of the quadrilaterals $A'B'QR$ and $ABQ'R'$ follows the equality of the segment QR and $Q'R'$ as well as the fact that $Q'R'$ is perpendicular to b. From this in turn the congruence of the quadrilaterals $QRMN$ and $Q'R'MN$ is concluded and thus the stated assertion, and at the same time Lemma 2 are completely proved.

LEMMA 3. Given any two nonparallel rays, there exists a line that is parallel to both of them, i.e., there exists a line which has the two prescribed ends α and β.

PROOF. Through any point O draw a parallel to the given ray and from O lay off equal segments, say up to A and B, so that

$$OA = OB$$

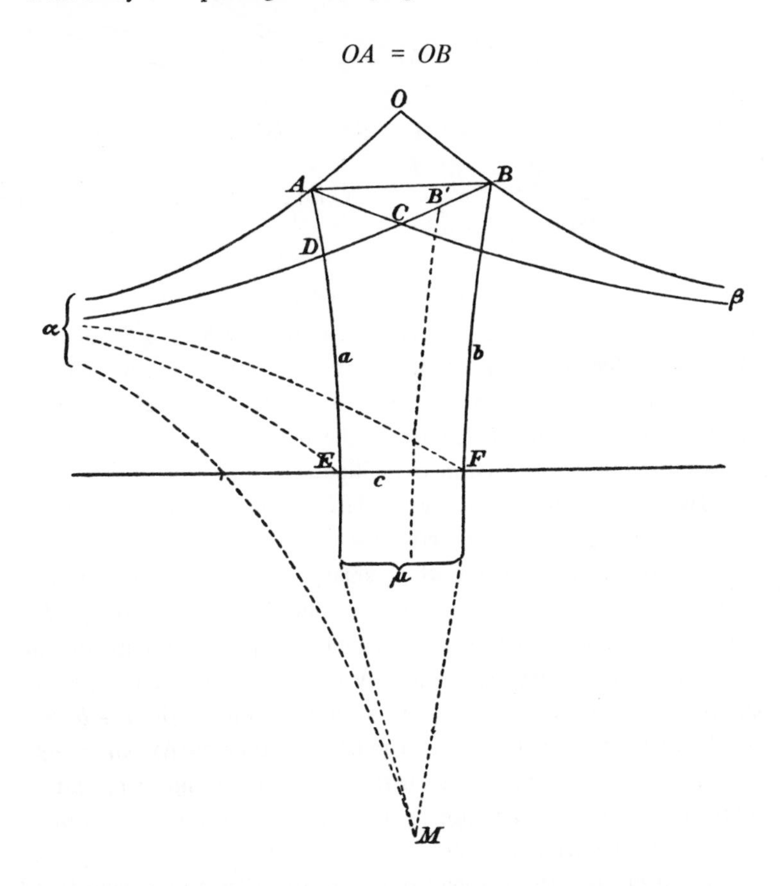

and the ray from O passing through A have the end α and the ray from O passing through B have the end β. Then connect the point A with the end β and bisect the angle between the two rays emanating from A. In like manner join the point B with the end α and bisect the angle between the two rays emanating from B. Let the first and the second bisecting lines be denoted by a and by b, respectively. From the congruence of the figures $OA\beta$ and $OB\alpha$ follows the equality of the angles

$$\measuredangle(OA\beta) = \measuredangle(OB\alpha),$$
$$\measuredangle(\alpha A\beta) = \measuredangle(\alpha B\beta),$$

and from the last equation the equality of the angles which have resulted from the bisection is also deduced, namely,

$$\measuredangle(\alpha A a) = \measuredangle(a A \beta) = \measuredangle(\alpha B b) = \measuredangle(b B \beta).$$

First it is necessary to show that the two bisecting lines a and b either intersect or are parallel.

Suppose a and b intersect at the point M. Since by construction OAB is an isosceles triangle it follows that

$$\measuredangle BAO = \measuredangle ABO$$

and hence by the preceding equations

$$\measuredangle BAM = \measuredangle ABM;$$

and thus

$$AM = BM.$$

Now connecting M with end α by a ray, the congruence of the figures αAM and αBM follows from the last segment equation and by the equality of the angles $\measuredangle(\alpha AM)$ and $\measuredangle(\alpha BM)$. This congruence would entail the equality of the angles $\measuredangle(\alpha MA)$ and $\measuredangle(\alpha MB)$. Since this conclusion is clearly invalid the assumption that the bisecting lines a and b intersect must be dropped.

Assume then that the lines a and b are parallel. Let the end determined by them be denoted by μ. Suppose that the ray emanating from B and passing through α meets the ray emanating from A and passing through β at the point C and the line a at the point D. The segments DA and DB will be shown to be equal. In fact, if this is not the case lay off DA on DB from D to, say, B' and connect B' with μ by a ray. From the congruence of the figures $DA\alpha$ and $DB'\mu$ would follow the equality of the angles $\measuredangle(DA\alpha)$ and $\measuredangle(DB'\mu)$, and thus the angles $\measuredangle(DB'\mu)$ and $\measuredangle(DB\mu)$, would be equal, which is impossible by Lemma 1.

The equality of the segment DA and DB yields now the equality of the angles $\measuredangle(DAB)$ and $\measuredangle(DBA)$ and hence by the foregoing the angles $\measuredangle(CAB)$ and $\measuredangle(CBA)$ are also equal and thus the equality of the angles $\measuredangle (DAB)$ and $\measuredangle (CAB)$ also follows. However this conclusion is evidently not true and thus this assumption, that lines d and b are parallel, must also be dropped.

Since according to these developments the lines a, b neither intersect nor are parallel there exists by Lemma 2 a line c that is perpendicular to both a and b, say at the points E and F, respectively. I assert that c is the desired line which connects the two given ends α and β.

For the proof assume to the contrary, that c does not have the end α. Then connect by rays each of the feet E and F with the end α. Connecting the midpoints of the segments AB and EF with each other it is easily seen that $EA = FB$. From this follows the congruence of the figures αEA and αFB and from it the equality of the angles $\measuredangle (AE\alpha)$ and $\measuredangle (BF\alpha)$, and thus the angles that are formed by the rays emanating from E and F with the line c are equal. This conclusion contradicts Lemma 1. Analogously, it can be shown that c also has the end β. The proof of the assertion is thus complete.

LEMMA 4. Let a, b be two parallel lines and O a point in the interior of the region of the plane lying between a and b. Furthermore let O_a be the reflected image of the point O in a, O_b the reflection image of the point O in b and M the midpoint of the segment O_aO_b. Then the ray from M that is parallel both to a and b is perpendicular to O_aO_b at M.

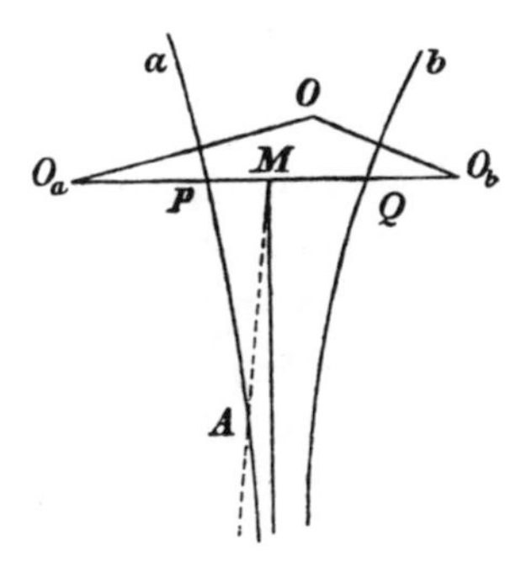

PROOF. If this is not the case, erect a perpendicular to O_aO_b at M on the same side of O_aO_b. Let the line O_aO_b intersect the lines a and b at the points P and Q, respectively. Since $PO < PQ + QO$ then $PO_a < PO_b$ and $QO_b < QO_a$. Thus M must lie in the interior of the region of the plane between a and b. The perpendicular at M must then meet the line a or b. If it meets a, at point A, then it would follow that $AO_a = AO$ and $AO_a = AO_b$, and consequently it would also follow that $AO = AO_b$, i.e., A could also be a point of b, which would contradict the hypothesis of the theorem.*

LEMMA 5. If a, b, c are three lines which have the same end ω, and if reflections in these lines be denoted by S_a, S_b, S_c, respectively, then there exists a line d, with the same end ω, such that the result of the consecutive applications of reflections in the lines a, b, c is the same as that of the reflection in the line d. This will be expressed by the formula

$$S_cS_bS_a = S_d.$$

PROOF. First assume that the line b lies in the interior of the region of the plane between a and c. Then let O be a point on b, and let the reflected images of O be denoted by O_a and O_c. Denoting now by d the line that connects the midpoint of the segment O_aO_c with the end ω, then in view of Lemma 4 the points O_a and O_c are reflection images in d and thus the operation $S_d\,S_c\,S_b\,S_a$ is such that it leaves the point O_a as well as the line that connects O_a with the end ω unchanged. Since this operation consists of four reflections the congruence theorems show that it is the identity. Hence follows the assertion.

Next, the validity of Theorem 5 can readily be seen in the case when the lines c and a coincide with each other. Indeed if b' is the line that results from a reflection of b in the line a then denoting by $S_b{}'$ the reflection in b' the validity of the formula

$$S_aS_bS_a = S_b{}'$$

can be seen immediately.

Finally, assume that the line c lies in the interior of the region of the plane between a and b. Then, by the first part of the proof there certainly exists a line d' such that the formula

$$S_aS_cS_b = S_d{}'$$

is valid. Denoting by d the reflection image of d' in a, then by the

*This conclusion agrees in essence with a result of Lobachevsky. Cf. Nicholas Lobachevsky, *Geometrical Researches on the Theory of Parallels*, trans. by G. B. Halstead (La Salle, Ill.: The Open Court Publishing Co., 1892 and 1914) Section 111. (Translator's note)

second part of the proof

$$S_c S_b S_a = S_a S_a S_c S_b S_a = S_a S_{d'} S_a = S_d.$$

Theorem 5 is thus completely proved.

§ 2. Addition of Ends

Consider a fixed line and denote its ends by 0 and ∞. Choose on this line $(0, \infty)$ a point O and erect a perpendicular at O. Let the ends of this perpendicular be denoted by $+1$ and -1.

The sum of two ends will be defined now as follows:

DEFINITION. Let α, β be any two ends distinct from ∞. Furthermore let O_α be the reflection image of the point O in the line (α, ∞) and let O_β be the reflection image of the point O in the line (β, ∞). Connect the midpoint of the segment $O_\alpha O_\beta$ with the end ∞. The other end of the line constructed in this manner will be called the *sum of the two ends* α and β and will be denoted by $\alpha + \beta$.

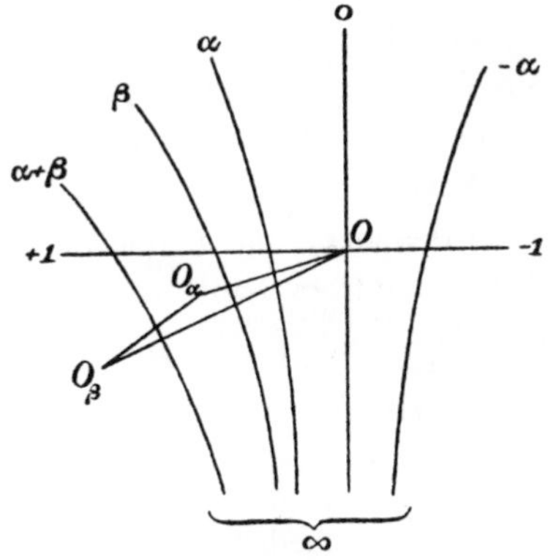

If a ray with the end α is reflected in the line $(0, \infty)$, the end of the resulting ray will be denoted by $-\alpha$.

It is easy to see the validity of the equations

$$\alpha + 0 = \alpha,$$
$$1 + (-1) = 0,$$
$$\alpha + (-\alpha) = 0,$$
$$\alpha + \beta = \beta + \alpha.$$

The last equation expresses **the commutative law of addition of ends.**

In order to show **the associative law of addition of ends** denote by S_0, S_α, S_β the reflections in the lines $(0, \infty)$, (α, ∞), (β, ∞), respectively. By Theorem 5 of Section 1 there exists a line (σ, ∞) such that the formula

$$S_\sigma = S_\beta S_0 S_\alpha$$

holds for the reflection S_σ in this line. Since in the operations $S_\beta S_0 S_\alpha$ the point O_α is transformed into the point O_β the latter must be the reflection image of the point O_α in the line (σ, ∞) and hence $\sigma = \alpha + \beta$, i.e., the formula

$$S_{\alpha + \beta} = S_\beta S_0 S_\alpha$$

holds.

If γ also denotes an end then the repeated application of the previously derived formula shows that

$$S_{\alpha + (\beta + \gamma)} = S_{\beta + \gamma} S_0 S_\alpha = S_\gamma S_0 S_\beta S_0 S_\alpha,$$
$$S_{(\alpha + \beta) + \gamma} = S_\gamma S_0 S_{\alpha + \beta} = S_\gamma S_0 S_\beta S_0 S_\alpha,$$

and hence

$$S_{\alpha + (\beta + \gamma)} = S_{(\alpha + \beta) + \gamma}$$

and thus also

$$\alpha + (\beta + \gamma) = (\alpha + \beta) + \gamma.$$

The previously derived formula

$$S_{\alpha + \beta} = S_\beta S_0 S_\alpha$$

shows at the same time that the given construction of the sum of two ends is independent of the choice of the point O on the line $(0, \infty)$. Hence if O' denotes any point distinct from O of the line $(0, \infty)$ and if O'_α, O'_β are the images of the point O' in the lines (α, ∞) and (β, ∞), respectively, then the perpendicular at the midpoint of $O'_\alpha O'_\beta$ is again the line $(\alpha + \beta, \infty)$.

Another result that is necessary for the development in Section 4 will be introduced here.

If the line (α, ∞) is reflected in the line (β, ∞) the resulting line is $(2\beta - \alpha, \infty)$.

In fact, if P is any point of the line which results from reflecting (α, ∞) in (β, ∞) then clearly it remains fixed if the reflections

$$S_\beta, S_0, S_{-\alpha}, S_0, S_\beta$$

are applied to it consecutively. However, in view of the above formula

$$S_\beta S_0 S_{-\alpha} S_0 S_\beta = S_{2\beta - \alpha},$$

i.e., the composite operation is equivalent to a reflection in the line $(2\beta - \alpha, \infty)$. Hence the point P must lie on the last line.

§ 3. Multiplication of Ends

The product of two ends will be defined as follows:

DEFINITION. If an end lies on the same side of the line $(0, \infty)$ as the end $+1$ does it will be said to be *positive*, and if an end lies on the same side of the line $(0, \infty)$, as the end -1 does, it will be said to be *negative*.

Let now α, β be any two ends distinct from 0 and ∞. Both lines $(\alpha, -\alpha)$ and $(\beta, -\beta)$ are perpendicular to the line $(0, \infty)$. Let these intersect this line at A and B, respectively. Now lay off on the line $(0, \infty)$ the segment OA from B to C in such a way that the direction on the line $(0, \infty)$ from O to A is the same as that from B to C. Then construct a perpendicular to the line $(0, \infty)$ at C and call the positive or negative end of this perpendicular the *product* $\alpha\beta$ *of the two ends* α, β according as both of these ends are positive, **or** negative, **or** one is positive and the other negative, respectively.

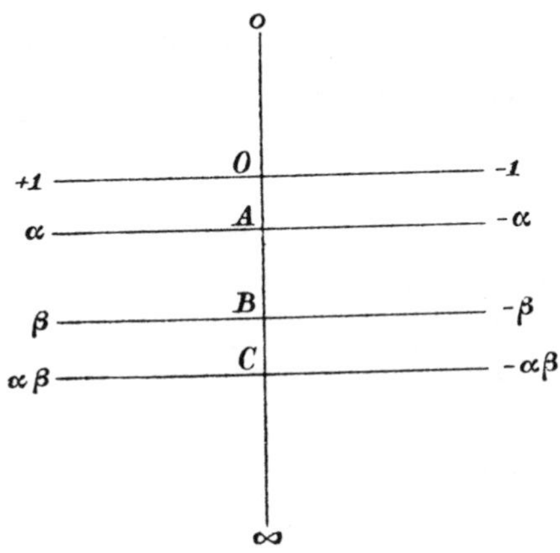

Finally assume the formula

$$\alpha \cdot 0 = 0 \cdot \alpha = 0.$$

By the triangle congruence axioms III, 1-3, the validity of the formulas

$$\alpha\beta = \beta\alpha,$$
$$\alpha(\beta\gamma) = (\alpha\beta)\gamma,$$

are immediately seen, i.e., the **commutative** as well as **the associative laws of multiplication of ends** hold.

It is also easily found that the formulas

$$1 \cdot \alpha = \alpha, \qquad (-1)\alpha = -\alpha$$

hold and that if the ends α, β of a line satisfy the equation

$$\alpha\beta = -1$$

then it must pass through the point O.

The possibility of division becomes clear immediately. Moreover, for every positive end π there always exists a positive (and also a negative) end whose square is equal to the end π and which therefore could be denoted by $\sqrt{\pi}$.

In order to prove **distributive law of the arithmetic of ends** first construct from the ends β and γ the end $\beta + \gamma$ by the method given in Section 2. Attempting then in the above indicated manner to determine the ends $\alpha\beta$, $\alpha\gamma$, $\alpha(\beta + \gamma)$, it can be seen that the construction is equivalent to the congruent mapping of the plane onto itself which produces a translation along the line $(0, \infty)$ by the segment OA.

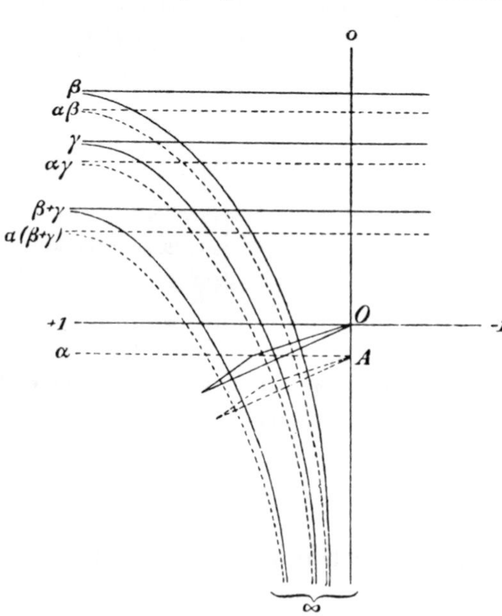

Therefore, if the sum of the ends $\alpha\beta$ and $\alpha\gamma$ is obtained by a construction from the point A instead of from the point O, which by one of the remarks in Section 2 is permissible, then indeed for this sum the end $\alpha(\beta + \gamma)$ is obtained, i.e., the formula

$$\alpha\beta + \alpha\gamma = \alpha(\beta + \gamma)$$

is valid.

§ 4. The Equation of a Point

Having seen in Section 2-3 that for the arithmetic of ends the same rules hold as for ordinary numbers, the construction of the geometry poses no further difficulties. It can be performed in the following way:

If ξ, η are the ends of any line let the ends

$$u = \xi\eta,$$

$$v = \frac{\xi + \eta}{2}$$

be called the *coordinates* of that line. The following fundamental proposition holds:

If α, β, γ are three ends with the property that the end $4\alpha\gamma - \beta^2$ is positive then all lines whose coordinates u, v satisfy the equation

$$\alpha u + \beta v + \gamma = 0$$

pass through the same point.

PROOF. Constructing the ends

$$\chi = \frac{2\alpha}{\sqrt{4\alpha\gamma - \beta^2}}, \qquad \lambda = \frac{\beta}{\sqrt{4\alpha\gamma - \beta^2}},$$

according to Section 2-3, then in view of the meaning of the coordinates u, v and since in case $\alpha \neq 0$, the given equation takes the form

$$(\chi\xi + \lambda)\ (\chi\eta + \lambda) = -1.$$

Let the transformation of an arbitrary variable end ω which is induced by the formula

$$\omega' = \chi\omega + \lambda,$$

be investigated. To do this consider first the transformations

$$\omega' = \chi\omega \quad \text{and} \quad \omega' = \omega + \lambda.$$

As far as the first transformation is concerned it is clear that multiplication of the arbitrary end ω by a constant χ is equivalent, according to Section 3, to a translation of the plane along the line $(0, \infty)$ by a segment depending on χ.

However, to the last transformation also i.e., to the addition of the end λ to the arbitary variable end ω, corresponds a certain motion of the plane on itself that depends only on λ, namely, one that can be regarded as a rotation of the plane about the end ∞.

In order to see this suppose that according to the discussion at the end of Section 2 through a reflection in the line $(0, \infty)$ the line (ω, ∞) is transformed into the line $(-\omega, \infty)$ and that it in turn goes through a reflection in the line $(\frac{\lambda}{2}, \infty)$ into the line $(\omega + \lambda, \infty)$ i.e., the addition of the end λ to the arbitrary variable end ω is equivalent to consecutive reflections in the lines $(0, \infty)$ and $(\frac{\lambda}{2}, \infty)$.

From what has been proved above it follows that if ξ, η are the ends of a line, then the ends of the line that is generated from it by a motion of the plane that depends only on χ and λ is given by the formulas

$$\xi' = \chi\xi + \lambda,$$
$$\eta' = \chi\eta + \lambda.$$

However, since the above equation

$$(\chi\xi + \lambda) \quad (\chi\eta + \lambda) = -1$$

yields the equation

$$\xi'\eta' = -1$$

for the ends ξ', η' and since according to one of the remarks of Section 3 this relation is the condition for the given lines to pass through the point O, it is seen that all lines (ξ, η) satisfying the original equation

$$(\chi\xi + \lambda) \quad (\chi\eta + \lambda) = -1$$

pass through one point. The proof of the stated theorem is thus complete.

Having seen that the equation of a point in line coordinates is linear it is easy to deduce the special case of Pascal's Theorem for a pair of lines and Desargues' Theorem for perspectively situated triangles as well as the other theorems of projective geometry. The familiar formulas of Bolyai-Lobachevskian geometry can then also be derived with no difficulty and the development of this geometry has been thus completed with the aid of Axioms I-IV alone.[1]

[1] As additional references to those at the beginning of this Appendix (on page 133, footnote 2) the following recent texts are cited:

F. Bachmann, *Aufbau der Geometrie aus dem Spiegelungsbegriff* (Berlin-Göttingen-Heidelberg, 1959).

K. Borsuk and W. Szmielew, *Podstawy Geometrii*, translated into English as *Foundations of Geometry*, by E. Marquit (Amsterdam, 1960).

The following articles might also be mentioned: W. Pejas, "Models of Hilbertian Axiom Systems of Absolute Geometry," *Math. Ann.*, Vol. 143 (1961), pp. 212-35; F. Bachmann, "Concerning Questions of Parallels," *Abh. Math. Seminar*, University of Hamburg, Vol. 27 (1964), pp. 173-92.

APPENDIX IV

FOUNDATIONS OF GEOMETRY[1]

(from *Math. Ann.*, Vol. 56, 1902)

Riemann's and Helmholtz's investigations of the foundations of geometry prompted Lie to undertake the axiomatic treatment of geometry by starting out with the group concept and led this brilliant mathematician to a set of axioms which he has shown with the aid of his transformation groups to be sufficient for the development of geometry.[2]

In developing his theory of transformation groups, Lie always assumed that the functions defining the groups can be differentiated, and thus in Lie's development there is no discussion of the question whether the assumption of differentiability as far as the geometric axioms are concerned is really indispensable, or whether the differentiability of the functions dealt with is rather a direct consequence of the group concept and the other geometric concepts. Because of his method Lie was obliged to formulate explicitly the axiom that the group of motions is generated by infinitesimal transformations. These requirements, as well as other essential parts of the axioms assumed by Lie with respect to the nature of the equation which defines equidistant points, can be expressed purely geometrically only by brute force and in a complicated way. Besides, they appear only through the analytic method used by Lie and are not due to the problem itself.

Therefore, I have attempted in what follows to formulate a set of axioms for plane geometry which while resting on the concept of a group contain only simple and geometrically clear requirements, and in particular assume in no way the differentiability of the functions induced by the motions. The axioms of the set formulated by me are contained as particular parts in the Lie axioms or, as I believe, can be derived immediately from them.

[1] For a characterization of the following development scheme of geometry, as compared to the schemes followed in the main part of this book, see the remark at the end of this article (p. 190).

[2] Lie (in collaboration with Engel), "Theorie der Transformationsgruppen," Vol. 3 (Leipzig, 1893), Section 5.

My line of reasoning is completely different from that of Lie's method. I work mainly with the concepts of point set theory developed by G. Cantor and have used Jordan's theorem according to which every closed continuous plane curve with no double points partitions the plane into an interior and an exterior region.

In the set formulated by me there are undoubtedly some redundancies. Nevertheless, I dispensed with any further investigation of this possibility in deference to the simple form of the axioms and above all because I wanted to avoid relatively complicated and geometrically unclear proofs.

In the following I discuss only the axioms for the plane, although I think that an analogous set of axioms for space, which makes an analogous development of space geometry possible,[1] can be formulated.

Some preliminary definitions will be given now.

DEFINITION. By the *number plane* is to be understood the ordinary plane with an orthogonal coordinate system x, y.

In this number plane a continuous curve without double points and including its end points will be called a *Jordan curve*. If a Jordan curve is closed then the **interior** of the region of the number plane bounded by it will be called a *Jordan region*.

For the sake of easier representation and comprehension I shall state the definition of the plane in the present investigation in a form

[1] I believe that through the following investigation it is possible to answer at the same time a general question concerning group theory, which I raised in my article "Mathematische Probleme," *Göttinger Nachrichten*, 1900, Problem 5, for the special case of motions in the plane.

that is narrower than my arguments require,[1] namely, I shall assume that it is possible to map all points of this geometry in a one-to-one manner onto the points lying in the finite part of the number plane as well as onto a fixed subset thereof so that every point of this geometry is characterized by a fixed pair of numbers x, y. This form of the concept of the plane is formulated as follows:

DEFINITION OF THE PLANE. *The plane is a set of objects called points which can be mapped on the points lying in the finite part of the number plane or some subset thereof in such a way that they have unique inverses.* These points of the number plane (i.e., the images) will also be used to denote the points of the plane.*

For every point A of the plane there exist in the number plane Jordan regions wherein lies the image of A and all of whose points also

[1] Regarding the broader form of the concept of the plane, compare my note on the foundations of geometry in *Göttinger Nachrichten*, 1902. There I have formulated the following more general definition of the plane:

The plane is a set of objects called points. Every point A determines certain subsets of points in which it is itself and which are called neighborhoods of the point.

It is always possible to map the points of a neighborhood on some Jordan region of the number plane in such a way that they have unique inverses. The Jordan region is called the image of that neighborhood.

Every Jordan region contained in an image inside of which lies the image of A is again an image of a neighborhood of A. Given distinct images of a neighborhood then the transformation with the unique inverses of one Jordan region on another induced by them is continuous.

If B is any point in a neighborhood of A then this neighborhood is also a neighborhood of B.

For every two neighborhoods of a point A there exists a third neighborhood of A that is common to the first two.

If A and B are any two points of the plane then there always exists a neighborhood of A which also contains the point B.

As it appears to me, for the case of two dimensions, these requirements contain the sharp definition of the concept which Riemann and Helmholtz called "multiply extended manifold" and which Lie called "number manifold" and on which they based all their investigations. They can also serve as the foundations for a rigorous axiomatic development of topology.

By assuming the above narrow definition of the plane, elliptic geometry is clearly excluded at the outset since its points cannot be mapped on the points lying in the finite part of the number plane in a manner that is consistent with one of the axioms. However, it is not difficult to recognize the changes necessary in the arguments if one assumes the broader form of the plane concept.

*As invertible mappings (transformations, functions) Hilbert considers also those which may not be unique or single valued, i.e., mappings that are not necessarily one-to-one and onto (cf. p. 193). (Translator's note)

represent points of the plane. These Jordan regions are called neighborhoods of the point A.

Every Jordan region contained in a neighborhood of A inside of which lies the point A (image of A) is again a neighborhood of A.

If B is any point in a neighborhood of A then it is at the same time also a neighborhood of B.

If A and B are any points of the plane then there exists a neighborhood of A which also contains the point B.

A motion will be defined as a single valued and invertible transformation of the plane on itself. Clearly two types of single valued and invertible transformations of the number plane can be distinguished at the very beginning. Taking any closed Jordan curve in the number plane and considering it to be oriented in a definite sense then by such a transformation it goes into a closed Jordan curve again which is oriented in a certain sense. In this investigation it is to be assumed that this orientation is the same as that of the original Jordan curve whenever a motion defining transformation of the number plane onto itself is applied. This assumption[1] implies the following form of the concept of a motion:

DEFINITION OF A MOTION. *A motion is a continuous transformation with unique inverses of the images of the number plane on itself such that the orientation of a closed Jordan curve always remains the same. The inverse of a transformation corresponding to a motion is again a motion.*

A motion in which a point M remains fixed is called a rotation about the point M.

Having established the concepts of "plane" and "motion," the following three axioms are formulated:

AXIOM I. *If two motions are performed consecutively then the resulting transformation of the plane onto itself is a motion again.*

Briefly stated:

AXIOM I. **The motions form a group.**

[1] This assumption is contained in Lie's requirement that the group of motions be generated by infinitesimal transformations. The opposite assumption (i.e., the assumption of the possibility of inversions) would essentially facilitate the proofs, inasmuch as it would be possible to define then immediately the "true line" as the locus of the points which remain fixed under an orientation changing transformation that leaves two points invariant.

AXIOM II. *If A and M are any distinct points of the plane then it is possible to bring the point A into an infinite number of positions by rotations about M.*

If the totality of points which are generated by all rotations about a point M from a point distinct from M is called a *true circle*[1] in this plane geometry then the assertion of Axiom II can be put in the following form:

AXIOM II. **Every true circle consists of an infinite number of points.**

The last required axiom will be preceded by a definition.

DEFINITION. Let AB be a definite pair of points in this geometry. Let the images of these points in the number plane be denoted by the same letters. Delineate neighborhoods α and β above the points A and B in the number plane, respectively. If a point A^* is in the neighborhood α and a point B^* is also in the neighborhood β it will be said that the pair of points A^*B^* lies in the neighborhood $\alpha\beta$ of AB. The expression that the neighborhood $\alpha\beta$ is as small as one pleases shall mean that α and β are respectively as small neighborhoods of A and B as one pleases.

Let ABC be a triple of points in this geometry. Let the images of these points in the number plane be denoted by the same letters. Delineate neighborhoods α, β, γ, about the points A, B, C in the number plane, respectively. If a point A^* is in the neighborhood α and a point B^* is in the neighborhood β and a point C^* is in the neighborhood γ it will be said that the triple of points $A^*B^*C^*$ is in the neighborhood $\alpha\beta\gamma$ of ABC. The expression that the neighborhood $\alpha\beta\gamma$ is as small as one pleases shall mean that α and β and γ are respectively as small neighborhoods of A and of B and of C as one pleases.

In the use of the phrases "pair of points" and "triple of points" it is not assumed that the points of the pair or the triple are necessarily distinct.

AXIOM III. *If there exist motions by which triples of points arbitrarily close to the triple $A\ B\ C$ can be carried arbitrarily close to*

[1] The expression "true circle" shall indicate that the figure defined in this manner in the course of the investigation shall prove to be isomorphic to the number circle. Corresponding interpretations hold for the expressions "true line" (p. 157) and "true segment" (p. 178).

the triple $A'B'C'$ then there exists a motion by which the triple $A\,B\,C$ is transformed precisely into the triple $A'B'C'$.[1]

The assertion of this axiom will be expressed briefly as follows:

AXIOM III. **The motions form a closed set.**

If in Axiom III some points of the triple are allowed to coincide, some special cases of Axiom III follow easily which are particulary noted as follows:

If there exist rotations about a point M by which pairs of points arbitrarily close to the pair AB can be carried arbitrarily close to the pair $A'B'$ then there always exists a rotation about M by which the pair AB is carried precisely into the pair $A'B'$.

If there exist motions by which pairs of points arbitrarily close to the pair AB can be carried arbitrarily close to the pair $A'B'$ then there exists a motion by which the pair AB is carried precisely into the pair $A'B'$.

If there exists rotation about the point M by which points arbitrarily close to the point A can be carried into points arbitrarily close to the point A' then there exists a rotation about M by which the point A is carried precisely into the point A'.

I shall often use the last special case of Axiom III in the subsequent arguments in such a way that the point M will take the place of A.[2]

I prove now the following assertion:

A plane geometry in which Axioms I - III are satisfied is either the Euclidean plane geometry or the Bolyai-Lobachevskian geometry.

If it is desired to obtain the Euclidean geometry alone then by Axiom I it is only necessary to add the provision that the group of motions shall contain a normal subgroup. This provision replaces the axiom of parallels.

---

[1] It is sufficient to assume that Axiom III is satisfied for sufficiently small neighborhoods, as is similarly assumed by Lie. My arguments can thus be modified so that only these narrower assumptions are used therein.

[2] A corollary which I stated as a special axiom in a lecture at the formal session of the anniversary celebration of the Gesellschaft der Wissenschaften zu Göttingen in 1901 is the following: "No two points can ever be brought arbitrarily close to each other by a motion." It remains to investigate to what extent or with which requirements, including this one, the above formulated Axiom III can be replaced.

I will outline briefly the train of thoughts of my arguments as follows:

By a special method a certain point figure *kk* is drawn in a neighborhood of any point M and on it a certain point K is constructed (Sections 1-2). The true circle χ through M and K, is then studied (Section 3). It follows then that this true circle χ is closed and dense in itself, i.e., it is a perfect point set.

The next objective of these developments is to show that the true circle χ is a closed Jordan curve.[1] This can be done by first demonstrating the possibility of ordering the points of the true circle χ (Sections 4-5), obtaining hence a mapping with unique inverses of the points of χ on the points of an ordinary circle (Sections 6-7) and finally showing that this mapping is necessarily continuous (Section 8). It follows then that the originally constructed point figure *kk* is identical with the true circle χ (Section 8). Then the theorem that every true circle inside χ is also a closed Jordan curve follows (Sections 10-12).

The investigation of the group of transformations by which the true circle χ goes into itself by rotating the plane about M is turned to next (Section 13). This group has the following properties: (1) Every rotation about M which leaves **one** point of χ fixed leaves all points of the latter fixed (Section 14). (2) There always exists a rotation about M which carries any given point of χ into any other point of χ (Section 15). (3) The group of rotations about M is continuous (Section 16). These three properties completely determine the structure of the group of transformations which corresponds to all rotations of the true circle onto itself. The following theorem is formulated: The group of transformations of the true circle χ onto itself which are rotations about M is holohedrally isomorphic to the group of ordinary rotations of the ordinary circle onto itself (Sections 17-18).

Next the group of transformations of **all** points of the plane by rotations about M is investigated. The theorem holds that except for the identity, there exists no rotation of the plane about M which leaves every point of a true circle μ fixed (Section 19). It is seen now that **every** true circle is a closed Jordan curve, and formulas for the

[1] Cf. to this an interesting note by A. Schönflies which pursues a similar objective, "Über einen grundlegenden Satz der Analysis Situs," *Göttinger Nachrichten*, 1902, and further expositions and references in *Berichte der Deutschen Mathematiker-Vereinigung*, supplementary Volume II (1908), pp. 158 and 178.

transformations of the group of all rotations about M are obtained (Sections 20-21). Finally, these theorems follow: If in a motion of the plane any two points remain fixed, then all points remain fixed, i.e., the motion is the identity. Every point of the plane can be carried by a suitable motion into every other point of the plane (Section 22).

The next important objective is to define the concept of a true line in this geometry, and to develop the necessary properties of this concept for the construction of this geometry. Next the concepts of a half rotation and the midpoint of a segment are defined (Section 23). A segment has at most one midpoint (Section 24), and if the midpoint of a segment is known it follows that every smaller segment also has a midpoint (Sections 25-26).

To examine the situation of the segment midpoints several theorems on tangent true circles are necessary and the first thing that comes to mind is the construction of two congruent circles which touch each other on the outside at one and only one point (Section 27). Next is derived a general theorem on circles which touch on the inside (Section 28), and then a theorem is deduced for the special case in which the circle touching on the inside passes through the center of the touched circle (Section 29).

Taking now as the unit segment some sufficiently small but definite segment, then by repeated bisections and half rotations thereof a set of points is constructed such that to every point of this set a definite number a becomes assigned, which is rational and has in the denominator only a power of 2 (Section 30). After formulating a rule for this assignment (Section 31), the points of the resulting set are ordered among themselves wherein the former theorems on tangent circles come into play (Section 32). Now it can be proved that the points corresponding to the numbers $1/2, 1/4, 1/8, \ldots$ converge to 0 (Section 33). This theorem is gradually generalized until it is seen that every sequence of points of this set converges as soon as the corresponding sequence of numbers converges (Sections 34-35).

Having made these preparations, one arrives at the definition of the true line as a set of points which is generated from two basis points, by repeatedly taking midpoints, performing half rotations, and adjoining all limit points of the resulting points (Section 36). Then it is possible to prove that the true line is a continuous curve (Section 37), which has no double point (Section 38), and has at most one point in common with any other true line (Section 39). It further follows that the true line intersects every circle drawn about any one of its points, and hence it follows that any two points of the plane can always be connected by a true line (Section 40). It is also seen that in this geometry, while the congruence theorems are valid, nevertheless two triangles are found to be congruent only if their orientations are the same (Section 41).

Regarding the relative positions of all true lines, there are two cases to be distinguished, depending on whether the axiom of parallels holds or whether for a given line there exist two lines through every point which separate the intersecting from the nonintersecting lines. In the first case Euclidean geometry, and in the second case Bolyai-Lobachevskian geometry, is arrived at (Section 42).

---

§ 1. Let M be any point in the geometry and also the image in the x, y number plane. The next objective is to construct certain point figures about M which eventually would turn out to be the true circles about M.

Describe in the number plane about M a "number circle," i.e., a circle **K** in the sense of ordinary metric length, which is so small that all points inside and on the circle **K** are also images and that there are also points outside **K**. Then there does exist inside **K** a circle **k** concentric with **K** such that all images of the circle **k** remain inside **K** by any rotation about M.

To prove this consider in the number plane an infinite sequence of concentric circles $\mathbf{k}_1, \mathbf{k}_2, \mathbf{k}_3, \ldots$ with decreasing radii which converge to 0. Assume contrary to the assertion that in every one of these circles there exists an image which by a certain rotation about M will lie outside the circle **K** or be moved onto its circumference. Let A_i be an image lying inside the circle $\mathbf{k}_i$ which goes outside the circle **K** or remains in the same position under the rotations Δ_i. Imagine then the radius r_i of every number circle $\mathbf{k}_i$ drawn from M to every one of the points A_i and consider the curve γ_i into which the radius r_i is transformed by the rotations Δ_i. Since the curve γ_i runs from the point M to some point outside or on the circle **K** it must meet the circumference of the circle **K**. Let B_i be such a point of intersection and B a limit point[1] of the points $B_1, B_2, B_3, \ldots$. Now let C_i be the

[1] By a *limit point* in this context is meant what today is customarily designated as an *accumulation point* [Häufungsstelle].

point on the radius r_i which is transformed by the rotation Δ_i into B_i. Since the points C_1, C_2, C_3, . . . converge to M there exists by Axiom III a rotation about M by which the point B lying on the circumference of the circle **K** goes into the point M. This contradicts the definition of a motion given above.

§ 2. As defined in Section 1 let **k** be a number circle inside **K** which satisfies the conditions of the theorem proved there so that all images inside **k** remain inside **K** under a rotation about M. Furthermore, let k be a number circle inside **k** all of whose points remain inside **k** under rotations about M. Let it be said briefly that the points of the number plane are *covered* if they result from the points inside or on k rotated in any way about M. From Axiom IV it follows immediately that the covered points form a closed set. Furthermore, let A be a definite point outside **K** whose image is a point of this geometry. Now let it be said that an uncovered point A' lies *outside kk* if it can be connected with A by a Jordan curve that consists of uncovered points alone. In particular, all points outside the number circle **k** surely lie outside kk. Let it be said that every covered point, every one of whose arbitrarily small neighborhoods contains points outside kk, is a point *on kk*. The points on kk form a closed set. Let the points J that are neither points outside kk nor points on kk be said to be points *inside kk*. In particular, all covered points not lying arbitrarily close to uncovered points, as, e.g., the point M and the points inside k, thus certainly lie inside kk.

§ 3. By noting, from the way **k** was determined, that under a rotation about M, A can never fall into the interior of **k**, it is seen that under every rotation about M the points outside kk go into points outside kk again, the points on kk go into points on kk again and the points inside kk go into points inside kk again.

According to the definition made, every point on kk is a covered point and since it is known that the points inside k also lie inside kk the following conclusion can be drawn:

For every point K on kk there exists a rotation Δ about M by which a point K' lying on the circumference of k falls on K. After the rotation Δ about M the radius MK' of the number circle k generates a Jordan curve which connects M with the point K on kk and otherwise lies entirely inside kk.

It is seen at the same time that at least one point of the circumference of the number circle k, namely, the point K', lies on kk.

Connect the point A that lies outside kk with M by any Jordan curve and denote now by K the point of this Jordan curve that lies on kk and which is such that all points on the Jordan curve between K and A lie outside kk. Consider then the set of all points generated from K by rotations about M, i.e., the true circle χ about M passing through K. All points of this true circle are on kk.

By Axiom II, χ contains infinitely many points. If K^* is a limit point of the points of the true circle χ then by Axiom III it is also a point of χ. Denoting any point of the true circle χ by K_1 then on performing the rotation about M which carries K^* into K_1 it follows that K_1 is also a limit point of the points of the true circle χ. The following theorem is thus obtained:

The true circle χ is a closed and dense set in itself, i.e., a perfect point set.

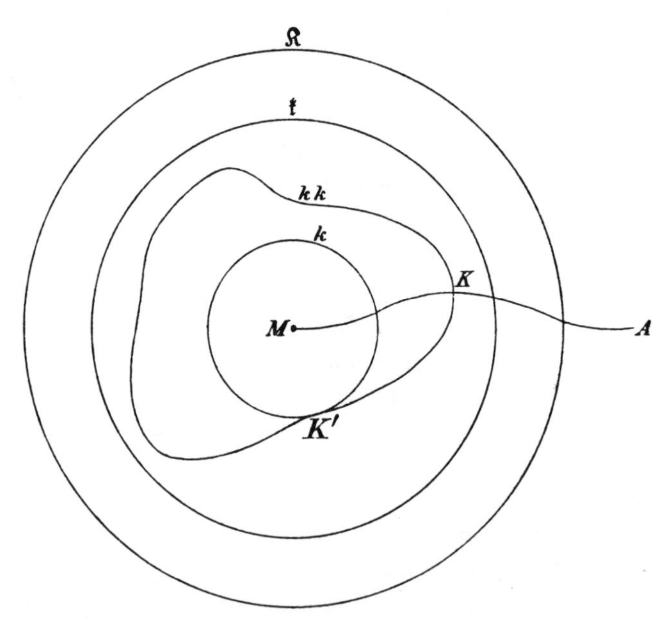

§ 4. The most important objective of the following developments is to show that the true circle χ is a closed Jordan curve. It will also turn out that the true circle χ coincides with kk.

First it will be shown *that any two points K_1, K_2 of the true circle χ can always be connected by a Jordan curve which except for its end points lies entirely inside kk as well as by a Jordan curve which except for its end points lies entirely outside kk.*

In fact, drawing the Jordan curves MK_1 and MK_2, in accordance with the above arguments, which connect inside kk the mid-point M with K_1 and K_2, respectively, and determining on the curve MK_1 originating at M the last point P that lies on MK_2, then the arc PK_1 of the first Jordan curve together with the arc PK_2 of the last Jordan curve form a connecting curve of the type originally desired.

On the other hand, consider the rotation about M by which K goes into K_1 or K_2. The point A_1 or A_2 which is thus generated from A, is (by Section 3), a point outside kk and can therefore be connected with A outside kk. From these connecting curves and the Jordan curves which are generated by these rotations from the Jordan curves constructed in Section 3, it is easy to form a Jordan curve between K_1 and K_2 which lies entirely outside kk.

§ 5. The theorem deduced above makes it possible to order the points of the true circle χ in a definite way.

Let K_1, K_2, K_3, K_4 be any four distinct points of the true circle χ. Connect the points K_1 and K_2 by a Jordan curve which (i.e., **between** K_1 and K_2) lies entirely inside kk, and by one which lies entirely outside kk. Since both of these connecting curves including their end points K_1, K_2 are continuous they form together a closed Jordan curve. Let a curve generated in this way from K_1, K_2 always be denoted by $\overline{K_1K_2}$. By the well-known Jordan curve theorem the entire number plane with the exception of $\overline{K_1K_2}$ is partitioned then into two regions, namely, into the interior and the exterior of the curve $\overline{K_1K_2}$. As far as the positions of the points K_3, K_4 are concerned, there are now two possibilities. The first is that the points K_3, K_4 are not separated by the curve $\overline{K_1K_2}$, i.e., both lie inside or outside it. The second is that the points K_3, K_4 are separated by $\overline{K_1K_2}$, i.e., K_3 lies inside and K_4 lies outside the curve $\overline{K_1K_2}$ or vice versa. If the points K_1, K_2 are connected in some other way by a path that lies entirely inside kk and by one that lies entirely outside kk then it is easy to see that for the positions of the points K_3, K_4, with respect to the new resulting closed Jordan curve $\overline{\overline{K_1K_2}}$, the same situation exists as before. Indeed, if for example the first case holds, and both points K_3, K_4 are in the interior of $\overline{K_1K_2}$, connect K_3 with K_4 by a path W lying inside kk. Should the path leave the interior of the closed curve $\overline{K_1K_2}$ then along the rest of its way it would eventually have to return to the interior. Hence it is certainly possible to replace the portion of the path W which lies outside $\overline{K_1K_2}$ by one that lies closer to the path of the arc of $\overline{K_1K_2}$

lying entirely inside *kk* and also inside $\overline{K_1K_2}$ so that there results a connecting path W^* between K_3 and K_4 which also lies entirely inside *kk* and inside $\overline{K_1K_2}$. Forming a new closed Jordan curve $\overline{\overline{\overline{K_1K_2}}}$ from the portion of the curve $\overline{K_1K_2}$ that lies inside *kk* and the portion of the curve $\overline{\overline{K_1K_2}}$ that lies outside *kk* then W^* is clearly a path which connects K_3 with K_4 inside this new Jordan curve without crossing the curve $\overline{\overline{\overline{K_1K_2}}}$, i.e., K_3 and K_4 are indeed not separated by $\overline{\overline{\overline{K_1K_2}}}$. Hence follows by a corresponding construction outside *kk* that neither are K_3 and K_4 separated by the curve $\overline{\overline{K_1K_2}}$. In the first case it may be simply said that the pair of points K_3, K_4 is not separated by the pair K_1, K_2. Then, however, it follows that also in the second case it may be simply said that the pair of points K_3, K_4 is separated by the pair K_1, K_2.

Perform now some rotation about M by which the points K_1, K_2, K_3, K_4 go into the points K'_1, K'_2, K'_3, K'_4. Noting that a rotation is by definition a continuous transformation with unique inverses of the number plane which carries points inside *kk* into points inside *kk* and points outside *kk* into points outside *kk*, it follows that the pairs of points K'_1, K'_2 and K'_3, K'_4 are or are not separated from each other according as the pairs of points K_1, K_2 and K_3, K_4 separate or do not separate each other, i.e., *the relative position of the pair of points* K_1, K_2 *and* K_3, K_4 *remains invariant under a rotation about* M.

The theorems that correspond to the other well-known facts about the relative positions of pairs of points on the circumference of an ordinary number circle can be derived in a similar way. These are the following:

If K_1, K_2 *are separated by* K_3, K_4 *then* K_3, K_4 *are separated by* K_1, K_2. *If* K_1, K_4 *are separated by* K_2, K_5 *and* K_2, K_4 *are separated by* K_3, K_5 *then* K_1, K_4 *are separated by* K_3, K_5.

This brings the following result:

The points of the true circle χ *are cyclic,* i.e., *they are ordered with respect to the relative separation of pairs of points just as the points of an ordinary number circle are. This ordering is invariant under the rotations of the true circle* χ *about the center* M.

§ 6. Another important property of the true circle χ is expressed as follows:

For every pair of points of the true circle χ *there always exists another pair on* χ *which separates the first pair.*

Denote by K_∞ a fixed point of the true circle χ and let it be said of any three points K_1, K_2, K_3 of the true circle χ that K_2 lies between K_1 and K_3 or not between K_1 and K_3 according as the pair of points K_1, K_3 is separated or not separated by the pair K_2, K_∞, respectively.

Let it be assumed, contrary to the above assertion, that K and K' are two points of the true circle χ which are not separated by any pair of points. It follows then by the definition made that no point of χ lies between them. Let it further be assumed that there exists no point K_1 such that the pair of points K_1, K' is separated by the pair K, K_∞. If this is not the case then let the roles of the points K and K' be interchanged in the following discussion. Choose then an infinite sequence R of points of the true circle χ which converge to the point K and connect K_1 with K' by a curve that lies inside kk as well by one that lies outside kk. By combining these two curves a closed Jordan curve $\overline{K_1K'}$ is obtained which separates K_∞ from K and therefore must also separate infinitely many points of R which converges to K. Let K_2 be one of these points of the sequence R. Since K_2 lies between K_1 and K' and cannot lie between K and K' then K_2 must lie between K_1 and K. Now connect in the same way K_2 with K' by a closed Jordan curve $\overline{K_2K'}$. A point K_3 of the sequence R is thus obtained which lies between K_3 and K, etc. One obtains in this way *an infinite sequence of points $K_1, K_2, K_3, \ldots$, each of which lies between its antecedent and K and which converge to the point K.*

Make now a rotation about M by which K goes into one of the points $K_1, K_2, K_3, \ldots$, say, K_i. Under this rotation let the point K' go into the point K'_i. Since by hypothesis K and K' are not separated by any pair of points the same holds for the pair K_i, K'_i. In view of this K'_i must coincide either with K_{i-1} or with K_{i+1} or lie between K_{i-1} and K_{i+1}. In any case, K'_i lies then between K_{i-2} and K_{i+2} so that the infinite sequence of points $K_1, K'_3, K_5, K'_7, K_9, K'_{11}, \ldots$ also has the property that every one of its points lying between its antecedent and the point K.

It will be shown now that the points $K'_3, K'_7, K'_{11}, \ldots$ must also converge to the point K. Indeed, if the points $K'_3, K'_7, K'_{11}, \ldots$ have a limit point Q distinct from K choose from among them a point K'_l. Since $K'_{l+4}, K'_{l+8}, K'_{l+12}, \ldots$ all lie between K'_l and K there exists a closed Jordan curve $\overline{K'_lK}$ which separates the point K_∞ from the points $K'_{l+4}, K'_{l+8}, K'_{l+12}, \ldots$ and therefore also from Q, i.e., Q must lie between K'_l and K. In view of the correspondence between the points K_i and K'_i it follows that Q also lies between the points K_1, K_5, K_9, . . . and K. The closed Jordan curve $\overline{QK_\infty}$ must therefore separate all points K_1, K_5, K_9, . . . from K. But then the points K_1, K_5, K_9, . . . cannot converge to K, as it should be.

Consider now the points K_3, K_7, K_{11}, . . . which converge to K and the points K'_3, K'_7, K'_{11}, . . . which according to what has been proved above converge at the same time to K. Since by a rotation about M the point K goes into K_i and at the same time K' goes into K'_i then by Axiom III there could exist a rotation which carries K and K' simultaneously into the common point of convergence K. This, however, would contradict the definition of a rotation. Thus, by refuting the assumption, the theorem stated at the beginning of this section has been completely proved.

§ 7. If by the definitions made at the beginning of Section 6 the true circle χ without the point K_∞ is considered as an ordered set of points in the sense of Cantor, *then the order type of this set is that of the line continuum.*

For the proof determine first a countable set S of points of the true circle whose limit points form the true circle χ itself. According to Cantor[1] the order type of such a set S is that of all rational numbers in their natural ordering, i.e., it is possible to make a correspondence between the points of the set S and the rational numbers in such a way that if A, B, C are any three points of S in which B lies between A and C then of the three assigned rational numbers a, b, c the number b by its value always lies between a and c.

Let now K be any point of the true circle χ which is not in the set S. Then if A, B are points of S they will be said to lie on different sides or on the same side of K accordingly as K lies or does not lie between A and B, respectively. Carrying over this definition for the points of S to their corresponding rational numbers one obtains a definite Dedekind cut of the set of rational numbers which is induced by the point K. Let the irrational number defined by this cut be made to correspond to the point K.

[1] "Beiträge zur Begründung der transfiniten Mengenlehre," *Math. Ann.*, Vol. 46, Section 9. For other conclusions in the text, compare Section 11 in particular. This memoir together with another one in *Math. Ann.*, Vol. 49, have been translated by Philip E. B. Jourdain, *Contributions to the Founding of the Theory of Transfinite Numbers* (La Salle, Ill.: The Open Court Publishing Co., 1911).

There cannot exist two distinct points K and K' on χ to which the same irrational number can correspond. Indeed, if a closed Jordan curve $\overline{KK'}$ is constructed and if H is any point of χ that lies between K and K', and consequently lies inside $\overline{KK'}$ then, since H is a limit point of the set S, there must also exist a point A in S that lies inside $\overline{KK'}$ and thus must also lie between K and K'. Consequently, the rational number a corresponding to A implies that the cuts induced by points K and K' are distinct.

It will finally be shown that, conversely, for every irrational number α there exists a point K on χ to which it corresponds. To do this let $a_1, a_2, a_3, \ldots$ be an increasing sequence of numbers and $b_1, b_2, b_3, \ldots$ a decreasing sequence of numbers, each converging to α. Construct the points $A_1, A_2, A_3, \ldots$ and $B_1, B_2, B_3, \ldots$ corresponding to these numbers and denote by K any limit point of these points $A_1, A_2, A_3, \ldots, B_1, B_2, B_3, \ldots$. The point K must correspond to the number α. For in general if a closed Jordan curve $\overline{A_iB_i}$ is constructed then the points $A_{i+1}, A_{i+2}, A_{i+3}, \ldots, B_{i+1}, B_{i+2}, B_{i+3}, \ldots$, and therefore also the limit point, will lie inside the curve $\overline{A_iB_i}$, i.e., between the points A_i, B_i. The cut induced by K is thus none other than the one that determines the point α.

Consider now the points on the circumference of any ordinary number circle of unit radius. Assign to one of these points the symbol $\pm\infty$ and also the point K_∞. To the remaining points, however, assign all real numbers in a continuous succession and to these in turn the corresponding points of the true circle χ. Then the following result is arrived at: *The points of the true circle χ can be orderly mapped on the points of the circumference of the ordinary number circle of unit radius, in such a way that they have unique inverses.*

§ 8. In order to attain the objective outlined in Section 4 it only remains to demonstrate the continuity of the mapping thus obtained, i.e., to show the absence of gaps in the true circle χ. To do this, suppose that the points of the true circle χ are determined by the coordinates x, y of the number plane and that the points of the number circle whose radius is 1 are determined by the arc t from a fixed reference point. Then it is necessary to show that x, y are continuous functions of t.

Now let $t_1, t_2, t_3, \ldots$ be any sequence of increasing or decreasing values converging to t^* and $K_1, K_2, K_3, \ldots$ the points of the true circle χ corresponding to these parametric values and let t^* correspond to a point K^* on χ. Furthermore, let Q be a limit point of the points $K_1, K_2, K_3, \ldots$. If a closed Jordan curve $\overline{K_iK^*}$ is constructed, the points $K_{i+1}, K_{i+2}, K_{i+3}, \ldots$, and therefore also their limit point Q, will necessarily lie inside K_iK^*, i.e., the point Q will also lie between K_i and K^*. Thus the parametric value t corresponding to Q would also have to be between t_i and t^*. The last contradiction could only be resolved if Q and K^* coincided. Consequently the points K_1, K_2, K_3, . . . converge to K^*. The continuity of the functions x, y with respect to the parameter t has thus been completely proved, and a result follows, which in Section 4 was stipulated as the first important objective of this development, namely, the following theorem:

The true circle χ *is a closed Jordan curve in the number plane.*

§ 9. It is known now that all points of the true circle χ are points on *kk*. It will turn out that the points of the latter all lie on χ, so that the following far-reaching theorem will hold:

The true circle χ *is identical with the points on kk. The points lying inside* χ *are also the points inside kk, and the points lying outside* χ *are also the points outside kk.*

To prove this theorem it will first be shown that the point M, the "center" of the true circle χ can be joined with every point J inside χ by a continuous curve without crossing the true circle χ.

Indeed, drawing an ordinary line in the number plane, a so-called "number line," through J, let K_1 and K_2 be the first points of this number line that lie on χ, counted in both directions from χ. Since K_1 and K_2 are also points on *kk* they can be connected with M by Jordan curves MK_1 and MK_2, respectively, both of which lie entirely inside *kk* and thus surely without crossing the true circle χ. If one of these Jordan curves meets the segment K_1K_2 at, say, the point B then the arc MB together with the segment JB form the desired connecting path. Otherwise, MK_1 and MK_2 together with the segment K_1K_2 form a closed Jordan curve γ. Since the curve γ lies entirely inside the number circle **k** (Section 1) the point A that lies outside the number circle **K** surely cannot be connected with a point inside γ without crossing over a point of the curve γ. The curve γ consists only of points inside *kk*, points on *kk* and points inside χ. Since the latter points from A on are accessible only by crossing over a point on χ which is also a point on *kk*, the region lying entirely inside γ must also lie inside *kk*. Connecting then M with the J by a continuous path lying inside γ it will surely not cross the true circle χ and is thus a path of the desired type.

It will be deduced from this that M lies inside χ, i.e., *the center M of the true circle χ lies inside the latter.*

Since every point of kk can also be connected by a Jordan curve which except for the end points lies entirely inside kk and thus surely does not meet χ, every point on kk must lie on χ or inside χ. If there existed a point P on kk which lay inside χ then the point A lying outside **K** could not be joined with points arbitrarily close to P without crossing over a point of χ. However, since every point of χ is covered, P could thus not be a point on kk. This is a contradiction. Consequently all points on kk lie also on χ whereby the above assertion is completely proved.

§ 10. The point figure kk was generated in Section 2 by a certain construction from the number circle k. Since, as shown in Section 3, the number circle k has at least one point on kk and otherwise lies entirely on or inside kk and by Section 9 the points on kk are none other than the true circle χ, that construction is also a means to form a true circle χ from the number circle k, the former being a closed Jordan curve surrounding the number circle which is tangent to it from the outside. Here, and in what is to follow, a Jordan curve which lies in the interior of another and has at least one common point with it will be said to be tangent to the other from the inside and the other will be said to be tangent to the first from the outside.

By a slight modification in the previous method, namely, by interchanging the roles assigned to the points inside and outside k another true circle can be constructed from the number circle k. Let it be said now that the points of the number plane are *covered* if they result from the points outside or on k rotated about M in any way and that all other points are *uncovered*. If an uncovered point can be connected with M by a Jordan curve that consists only of uncovered points, let it be said to be *inside kkk*. Let the boundary points of the points inside kkk be said to be points on kkk and all other points to be outside kkk. It will be shown, similarly in Sections 3-9, *that the points on kkk form a true circle about M which is a closed Jordan curve that surrounds the center M and lies inside the number circle k and is tangent to it from the inside.*

§ 11. Instead of the number circle k it is possible to choose now any closed Jordan curve z lying inside k which contains the point M in its interior. *By applying the same construction, one obtains for the curve z a definite true circle about M which surrounds it and which is a closed Jordan curve tangent to z from the outside as well as a definite true circle about M that lies inside z and which is a closed Jordan curve tangent to z from the inside.*

Note that every such true circle constructed from a Jordan curve z can also be constructed from a number circle. It is only necessary to choose a number circle which lies in the given true circle and which is tangent to it from the inside or one that surrounds it and is tangent to it from the outside. For two true circles that are closed Jordan curves, whether they both surround the same number circle or lie entirely inside it, and are tangent to it they must have a common point and hence are identical.

§ 12. It is possible to prove now without undue difficulty the important result *that every true circle about M passing through any point P inside* χ, *like the true circles constructed in Section 11, is a closed Jordan curve that contains M in its interior.*

For the proof consider on the one hand all true circles about M which are closed Jordan curves and which do not surround P. Let them be called true circles of the *first* kind. On the other hand consider all circles which are closed Jordan curves and which do surround P. Let them be called true circles of the *second* kind.

First, consider the *surrounding* true circle generated from every number circle whose center is M and examine the number circles from which true circles of the first kind arise. Then find for these number circles, the boundary circle g, i.e., the smallest number circle which contains all of them. All number circles that are smaller than g yield then true circles of the first kind. If the true circle γ arising from the number circle g did not pass through P it could not surround that point either. For if P lay inside γ then a closed Jordan curve could be drawn entirely inside γ surrounding the points M and P from which the true circle which it surrounded could be generated. This true circle, since it definitely lies in the interior of the number circle g, could be generated from a number circle that is smaller than g. Furthermore, it could surround the point P, which is impossible. Since, as noted, all true circles about M which are closed Jordan curves also arise from number circles about M it is clear that the true circle that arises from g is a true circle of the first kind that surrounds all other true circles of the first kind.

On the other hand, by considering the true circle generated from every number circle whose center is M and which it does not surround, the existence of a true circle of the second kind which is surrounded by all other true circles of the second kind can be shown in a similar way.

Now if the two true boundary circles thus found did not pass through P then a Jordan curve could be drawn in the annular region between them which, by the given method, would certainly yield a true circle that would be a closed Jordan curve, but neither of the first kind nor of the second kind. This is a contradiction and the assertion stated at the beginning of Section 12 is thus proved.

§ 13. Having found in the foregoing the most important properties of the true circles about M which pass through points inside χ, *the investigation of the group of all motions by whose rotations of the plane about M the true circle* χ *goes into itself* will be turned to next.

Let the points of the true circle χ be orderly mapped onto the points t of the circumference of a number circle of unit radius in accordance with the developments in Section 8. Then to every rotation Δ of the plane about M corresponds a definite continuous transformation with unique inverses of the points t of the unit circle onto itself, since by Section 5 the order of the points on the true circle and hence, according to Section 7, also the order of the parametic values t, remain invariant under a rotation. This transformation can be represented by a formula of the form

$$t' = \Delta(t).$$

where $\Delta(t)$ is a continuous function which monotonically increases or decreases with increasing t and which by an increase in the argument t by 2π also changes in value by 2π.

To the functions $\Delta(t)$ which decrease with increasing argument t correspond transformations which change the orientation of the true circle, and since by the definition adopted for a motion its orientation must remain the same, it follows that the function $\Delta(t)$ must always increase with increasing t.

§ 14. Let the question be posed now whether in the group of all rotations about M there can exist a rotation under which a point A of the true circle χ can remain invariant. Let $t = a$ be the parametric value of such a point A, and let it remain fixed under the proper rotation Δ, which is represented by the formula

$$t' = \Delta(t).$$

Furthermore, let B be any point of the true circle whose parametric value is $t = b$ and which changes its position under the rotation Δ. Assume, let us say, that $b < a$, whereby there is no loss of generality.

$\Delta(t)$ as well as its inverse $\Delta^{-1}(t)$ increases with increasing argument. In view of $\Delta(a)=a$ it can be concluded successively that all values which can be represented by the symbolic powers

$\Delta(b)$, $\Delta\Delta(b) = \Delta^2(b)$, $\Delta^3(b)$,. . ., $\Delta^{-1}(b)$, $\Delta^{-2}(b)$, $\Delta^{-3}(b)$, . . .

are all less than a. In case $\Delta(b) > b$ then the values

$$\Delta(b),\ \Delta^2(b),\ \Delta^3(b),\ \ldots$$

form a monotonically increasing sequence of numbers. In case $\Delta(b) < b$ then the same holds for the sequence of values

$$\Delta^{-1}(b),\ \Delta^{-2}(b),\ \Delta^{-3}(b),\ \ldots$$

From these facts one concludes that in the first case the successive applications of the rotation Δ to b and that in the last case the symbolic powers of $\Delta(b)$ with negative exponents both must approach a limiting value g which either is between a and b or coincides with a. If the limit g corresponds to some point G on the true circle χ then the powers of Δ with positive or negative exponents form motions by which the point B can eventually go into points that are arbitrarily close to G and by which at the same time points in arbitrarily small neighborhoods of G remain in arbitrarily small neighborhoods of G. Consequently by Axiom III there could exist a motion which carries B into G and at the same time leaves G invariant. This would contradict the definition of a motion. *Consequently the rotation Δ which leaves the point A fixed is necessarily one that leaves all points of the true circle χ fixed, i.e., for that circle it is the identity.*

§ 15. From the definition of the true circle the following fact is immediately apparent:

There always exists a rotation about M which carries any given point O of the true circle χ into any other given point S on it.

§ 16. Another property of the group of motions of a true circle onto itself will be deduced now.

Let O, S, T, Z be four points on the true circle χ such that the rotation about M by which O is carried into S move the point T towards Z so that the position of Z is uniquely determined by the points O, S, T. Holding *O fixed* and moving S and T on the true cirle *a continuous variation of S and T produces also a continuous variation of Z.*

To prove this, choose an infinite sequence of points S_1, S_2, S_3, . . . which converges to the point S and an infinite sequence of points T_1, T_2, T_3, . . . which converges to the point T. Denote by Δ_1, Δ_2, Δ_3, . . . the rotations about M by which O goes into S_1, S_2,

$S_3, \ldots$ and let the points which arise from $T_1, T_2, T_3, \ldots$ by the respective rotations $\Delta_1, \Delta_2, \Delta_3, \ldots$ be $Z_1, Z_2, Z_3, \ldots$, respectively. Then it is necessary to show that the points $Z_1, Z_2, Z_3, \ldots$ converge to Z. Let Z^* be a limit point of the points $Z_1, Z_2, Z_3, \ldots$. By Axiom III there exists then a rotation about M by which O goes into S and at the same time T goes into Z^*. The latter is thus seen to be uniquely determined and identical to Z.

§ 17. In Section 14-16 it has been found that the group of all rotations of the true circle χ onto itself has the following properties:

1. Except for the identity, there exists no rotation about M that leaves a point of the true circle χ fixed.

2. If O, S are any two points of the true circle χ, then there exists a rotation about M which carries O into S.

3. Under a rotation about M which moves O to S, T goes into Z at the same time. The uniquely determined point Z by O, S, T thus undergoes a continuous variation on χ if S and T change their positions on χ continuously.

These three properties determine completely the structure of the group of transformations $\Delta(t)$ which correspond to the motions of the true circle onto itself. The following theorem can then be stated:

The group of all motions of the true circle χ onto itself, which are rotations about M, is isomorphic to the group of the ordinary rotations of the number circle about M onto itself.

§ 18. Suppose that the rotation about M which carries the point O of the true circle χ with the parametric value 0 onto the point S with the parametric value s is represented by the transformation formula

$$t' = \Delta(t, s)$$

for which $\Delta(t, 0) = t$ is taken. Then by the properties found for the rotation group it is seen that the function $\Delta(t, s)$ is single valued and continuous in both variables t, s. It also follows that s is uniquely determined to within a multiple of 2π by two corresponding values t and t', that the function $\Delta(t, s)$ with constant t and increasing s either only increases or only decreases monotonically and since for $t = 0$ it becomes s, the first case must occur. Now

$$\Delta(t, t) > \Delta(0, t), \quad \Delta(0, t) = t; \quad (t > 0),$$

and in view of $\quad \Delta(2\pi, s) = 2\pi + \Delta(0, s) = 2\pi + s$

it follows that $\quad \Delta(2\pi, 2\pi) = 4\pi.$

Thus the function $\Delta(t, t)$ $(> t)$ of the single variable t has the property of increasing monotonically from 0 to 4π when the argument t increases from 0 to 2π. Hence the following is concluded immediately:

Given any positive number $t' \leqq 2\pi$ there exists one and only one positive number t such that

$$\Delta(t, t) = t'$$

with $t < t'$. The parametric value t yields a point of the true circle such that by a certain rotation about M the point $t = 0$ moves to t and at the same time the point t moves to t'.

Let the value t for which

$$\Delta(t, t) = 2\pi$$

be denoted now by $\varphi\left(\frac{1}{2}\right)$, and the one for which

$$\Delta(t, t) = \varphi\left(\frac{1}{2}\right)$$

by $\varphi\left(\frac{1}{2^2}\right)$, the one for which

$$\Delta(t, t) = \varphi\left(\frac{1}{2^2}\right)$$

by $\varphi\left(\frac{1}{2^3}\right)$, . . . ; Furthermore, let

$$\Delta\left(\varphi\left(\frac{a}{2^n}\right), \quad \varphi\left(\frac{1}{2^n}\right)\right) = \varphi\left(\frac{a+1}{2^n}\right),$$

where a is an integer and n is an integer that is greater or equal to 1, and further, let

$$\varphi(0) = 0, \quad \varphi(1) = 2\pi.$$

The function φ is thus consistently defined for all rational arguments whose denominators are powers of 2.

Now if σ is any positive argument that is less than 1 let it be expanded in a binary fraction of the form

$$\sigma = \frac{z_1}{2} + \frac{z_2}{2^2} + \frac{z_3}{2^3} + \cdots,$$

where all $z_1, z_2, z_3,$ . . . denote the digit 0 or 1. Since the numbers of the sequence

$$\varphi\left(\frac{z_1}{2}\right), \quad \varphi\left(\frac{z_1}{2} + \frac{z_2}{2^2}\right), \quad \varphi\left(\frac{z_1}{2} + \frac{z_2}{2^2} + \frac{z_3}{2^3}\right), \ldots$$

are nondecreasing and all remain less than or equal to φ (1) they approach a limit. Let it be denoted by $\varphi(\sigma)$. The function $\varphi(\sigma)$

increases monotonically with increasing argument. It will be shown that it is also continuous. Indeed, if it were discontinuous at a point

$$\sigma = \frac{z_1}{2} + \frac{z_2}{2^2} + \frac{z_3}{2^3} + \cdots = \underset{n=\infty}{L} \frac{a_n}{2^n} = \underset{n=\infty}{L} \frac{a_n + 1}{2^n},$$

$$\left(\frac{a_n}{2^n} = \frac{z_1}{2} + \frac{z_2}{2^2} + \cdots + \frac{z_n}{2^n}\right)$$

then the two limits

$$\underset{n=\infty}{L} \varphi\left(\frac{a_n}{2^n}\right) \quad \text{und} \quad \underset{n=\infty}{L} \varphi\left(\frac{a_n + 1}{2^n}\right)$$

could be different and thus the infinite sequence of points corresponding to the parameters

$$t = \varphi\left(\frac{a_1}{2}\right), \quad t = \varphi\left(\frac{a_2}{2^2}\right), \quad t = \varphi\left(\frac{a_3}{2^3}\right), \ldots$$

could converge to a point other than that to which the infinite sequence of points corresponding to the parameters

$$t = \varphi\left(\frac{a_1 + 1}{2}\right), \quad t = \varphi\left(\frac{a_2 + 1}{2^2}\right), \quad t = \varphi\left(\frac{a_3 + 1}{2^3}\right), \ldots$$

converges. Now the rotation which carries the point $t = \varphi\left(\frac{a_n}{2^n}\right)$ into the point $t = \varphi\left(\frac{a_n + 1}{2^n}\right)$ carries at the same time the point $t = \varphi\left(\frac{1}{2^n}\right)$ the point $t = \varphi\left(\frac{1}{2^n - 1}\right)$ and since the numbers $\varphi\left(\frac{1}{2}\right), \varphi\left(\frac{1}{2^2}\right), \varphi\left(\frac{1}{2^3}\right), \ldots.$ decrease monotonically and the parameters corresponding to these points must converge to a point A; then, by a frequently applied argument based on Axiom III, the preceding infinite sequences of points must also both converge to the same point.

Since the function $\varphi(\sigma)$ is monotonically increasing and continuous, it has a single valued and continuous inverse.

The rotation about M by which the point $t = 0$ goes into the point $t = \varphi\left(\frac{a_n}{2^n}\right)$ carries at the same time the point $t = \varphi\left(\frac{b_m}{2^m}\right)$ into $t = \varphi\left(\frac{b_m}{2^m} + \frac{a_n}{2^n}\right)$ where b_m is some integer. Since for $n = \infty$ the values $\varphi\left(\frac{a_n}{2^n}\right)$ converge to $\varphi(\sigma)$, and at the same time the numbers $\varphi\left(\frac{b_m}{2^m} + \frac{a_n}{2^n}\right)$ converge to $\varphi\left(\frac{b_m}{2^m} + \sigma\right)$ by Axiom III there exists a rotation which moves the point $t = 0$ to $t = \varphi(\sigma)$ and at the same time moves the point $t = \varphi\left(\frac{b_m}{2^m}\right)$ to $t = \varphi\left(\frac{b_m}{2^m} + \sigma\right)$, i.e.,

$$\Delta\left(\varphi\left(\frac{b_m}{2^m}\right), \quad \varphi(\sigma)\right) = \varphi\left(\frac{b_m}{2^m} + \sigma\right),$$

and since φ is a continuous function, it follows that for arbitrary parameters τ, σ

$$\Delta(\varphi(\tau),\ \varphi(\sigma)) \doteq \varphi(\tau + \sigma).$$

It has thus been shown that if new parameters τ, τ', σ are introduced for t, t', s in the transformation formula

$$t' = \Delta(t, s)$$

by means of some function φ having a unique inverse with

$$t = \varphi(\tau), \quad t' = \varphi(\tau'), \quad s = \varphi(\sigma)$$

then the rotation can be expressed in the new parameters by the formula

$$\tau' = \tau + \sigma$$

This theorem proves the validity of the assertion made in Section 17.

Substitute now the parameter $\omega = 2\pi\sigma$ for σ and let it be called *the angle* or *the arc length* between the point $O(\sigma = 0)$ and S (i.e., σ) on the true circle χ. Let the rotation by which the point O $(\sigma = 0)$ goes into the point S (i.e., σ) be called *a rotation* $\Delta\ [\omega]$ *of the true circle* χ *onto itself through the angle* ω.

§ 19. With the proof of the theorem in Section 17, the investigation of the rotations of the true circle χ onto itself is finished. In view of Sections 11 and 12 it can be seen that the conclusions applied to and results proved for the true circle χ are also valid for all true circles about M that lie inside χ.

The group of transformations of **all** points by the rotations of the plane about the fixed point M will be turned to next and the following theorems will be proved successively:

Let it be given that a true circle μ about M is a closed Jordan curve in whose interior lies M. Then except for the identity there exists no rotation of the plane about M which leaves every point of the true circle μ fixed.

For the proof let a rotation about M that leaves every point of μ fixed be denoted by **M** and as the *first* thing contrary to the assertion assume that there exist points arbitrarily close to a point A on μ which change their positions by the rotation **M**. Draw a true circle α about A which passes through one of the points changing under **M** that is sufficiently small so that by the above remark it satisfies the theorem in Section 14. By Section 12 it is possible to do this. Let B be a point of intersection of this circle with μ. Then the rotation **M** can immediately be characterized as a rotation of the circle α onto itself by which B remains fixed. However, by Section 14 all points on α remain fixed under such a rotation, which is not the case. The first assumption is thus seen to be untenable.

Construct now a set of closed Jordan curves about M, of which μ is one, and such that one either completely contains the other or completely surrounds the other, so that one and only one curve of the set passes through every point of the number plane. Now as the *second* thing contrary to the above assertion assume that in this set there exists a curve λ inside or outside μ such that all points in the annular region between μ and λ remain fixed under every rotation **M** whereas arbitrarily close to the curve λ there exist points which do not remain fixed under every rotation **M**.

Let A be a point on λ arbitrarily close to which lie points movable by **M**. Draw a true circle α about A which passes through one of these movable points and is sufficiently small so that the theorem in Section 14 is applicable to it. Since this circle, by being sufficiently small, passes in any case through part of the annular region which remains fixed under the motions **M**, the motion **M** can be characterized immediately as a rotation of the cirlce α onto itself under which infinitely many points of α remain fixed. However, by Section 14 all points of α could then remain fixed under **M**, which is not the case. This shows then that under all rotations **M** the points of the plane remain fixed.

§ 20. The following important assertion is stated now:

Every true circle is a closed Jordan curve. The set of all true circles about any point M fills the plane without gaps so every true circle about M either contains or surrounds every other true circle. The totality of rotations $\Delta[\omega]$ *of the plane about M can be expressed by transformation formulas of the form*

$$x' = f(x, y; \omega), \quad y' = g(x, y; \omega)$$

where x, y *and* x', y' *are coordinates in the number plane and* f, g *are single valued continuous functions in the three variables* x, y, ω. *Furthermore, the smallest simultaneous period of the functions* f, g *for every point* x, y, *with respect to the argument* ω *is* 2π, *i.e., every point of the true circle is obtained once and only once from the point* (x, y) *every time* ω *is allowed to range from* 0 *to* 2π. *Finally for the composition of two rotations through the angles* ω, ω' *the formula*

$$\Delta[\omega]\,\Delta[\omega'] = \Delta[\omega + \omega'].$$

holds.

§ 21. To prove the stated assertion, first consider again the true circle χ about M, investigated in Sections 3 through 18, which is a closed Jordan curve, and examine the rotations of this circle onto itself. Introduce the angle ω in accordance with Section 18 so that by specifying a value for ω between 0 and 2π a motion of the true circle χ on itself becomes uniquely determined. However, to every rotation of the true circle onto itself there corresponds only **one** definite rotation of the plane about M, since according to Section 19 all points of the plane as a whole remain fixed by keeping all points on χ fixed. Hence follows that the functions f, g given in the formulas in Section 20 for a rotation of the plane about M are *single valued* for all x, y, ω whose period with respect to ω is 2π.

It will be shown next that f, g are *continuous* functions in x, y, ω. To do this let O be any point on χ. Furthermore, let $\omega_1, \omega_2, \omega_3, \ldots$ be an infinite sequence of values which converges to a definite value ω, and T_1, T_2, T_3, ... an infinite sequence of points of the plane which converges to some point T. Let the points which are generated from O by applications of the rotations through the angles $\omega_1, \omega_2, \omega_3, \ldots$ be denoted by S_1, S_2, S_3, ..., and the points which are generated from T_1, T_2, T_3, ... by the rotations $\omega_1, \omega_2, \omega_3, \ldots$, respectively, let them be designated by Z_1, Z_2, Z_3, ..., respectively. Finally, let the points which are generated from O and T by a rotation through the angle ω be denoted by S and Z, respectively. It is sufficient then to show that the points Z_1, Z_2, Z_3, ... converge to Z.

Since the points T_1, T_2, T_3, ... converge to T it is possible to determine a Jordan region G in whose interior lie all the points M, T, T_1, T_2, T_3 .... Apply to this Jordan region a rotation about M which moves O to S. Let the Jordan region formed in this way from G be denoted by H. It contains the points M and Z. Finally construct a closed Jordan curve α which contains the region H entirely in its interior, i.e., it surrounds it with no point lying in H.

It will be shown that of the points Z_1, Z_2, Z_3, ... only a finite number lie outside the curve α. In fact suppose that an infinite number of them, say the points Z_{i_1}, Z_{i_2}, Z_{i_3}, ..., lie outside α. Then M can be connected by T_{i_h} by a Jordan curve γ_h lying inside G and a rotation can be performed with γ_h through the angle ω_{i_h}. The curve resulting from this connects M with Z_{i_h} and consequently intersects the curve α at some point, say B_h. Let A_h be the point on γ_h which goes into B_h by the rotation through the angle ω_{i_h}. Since all the points A_1, A_2,

$A_3, \ldots$ lie inside G and all the points $B_1, B_2, B_3, \ldots$ remain on α there exits an infinite sequence of indices $h_1, h_2, h_3, \ldots$ such that $A_{h_1}, A_{h_2}, A_{h_3}, \ldots$ converge to a point A inside G or on the boundary of G and at the same time $B_{h_1}, B_{h_2}, B_{h_3}, \ldots$ converge to a point B on α. Thus the points $S_1, S_2, S_3, \ldots$ converge to S. By Axiom III there could exist a rotation about M which moves O to S and at the same time A to B. This however would be impossible. For under such a rotation A must go into a point inside H or on the boundary of H. However, B is a point on the curve α which contains the region H entirely in its interior.

It has thus been seen that the set of points $Z_1, Z_2, Z_3, \ldots$ must lie entirely inside some Jordan region.

Let Z^* be now a limit point of the points $Z_1, Z_2, Z_3, \ldots$. Since the points $S_1, S_2, S_3, \ldots$ converge to S there exists by Axiom III a rotation about M under which O goes into S and at the same time T goes into Z^*. However, since under the rotation about M which carries O into S, T should go into Z it follows, in view of the single valuedness of the functions f, g proved above, that necessarily $Z^* = Z$, i.e., the points $Z_1, Z_2, Z_3, \ldots$ accumulate only at one point, namely at Z. The continuity of the functions f, g in x, y, ω has thus been proved.

Substitute now in f, g for x, y the coordinates of any point P of the plane that lies inside or outside the circle χ. The resulting functions $f(\omega)$, $g(\omega)$ in the variable ω alone cannot have arbitrary small simultaneous periods. For since they are continuous functions of ω they would then be constants. But then the point P would remain fixed under all rotations of the plane about M, which would contradict Axiom II. Consequently the smalles simultaneous period of the functions $f(\omega), g(\omega)$ must be of the form $\frac{2\pi}{n}$ where n is a positive integer. Hence follows that the true circle which passes through P can be obtained by letting ω range from 0 to $\frac{2\pi}{n}$ in the formulas

$$x = f(\omega), \quad y = g(\omega).$$

This curve is closed and has no double points. It thus represents the true circle passing through P. Applying then a rotation through the angle $\frac{2\pi}{n}$ to the plane, all points of the true circle through P remain fixed, and hence by Section 19 all points of the plane must then remain fixed. However, the points on the true circle χ remain fixed only if $n = 1$. The assertions of the theorem stated in Section 20 have thus been completely proved.

§ 22. It is easy to see now the validity of the following results:

If any two points of the plane remain fixed under a motion of the plane then all points remain fixed, i.e., the motion is the identity.

Every point of the plane can be carried into any other point of the plane by a motion (i.e., two rotations).

The first result follows immediately by the theorem in Section 20. The second follows from the fact that drawing the true circle about each of the points and through the other, the circles must meet.

§ 23. The next most important objective *is to introduce the concept of the true line into this geometry and to deduce the necessary properties for the development of this geometry.*

To this end the following terminology is introduced: If A, B and A', B' are two pairs of image points such that A can be carried into A' and B into B' simultaneously by one motion it will be said that *the (true) segment AB is congruent to* (symbolically $\equiv$) *the (true) segment* $A'B'$. Furthermore *two true circles* will be said to be *congruent* if there exists a motion which carries their centers and themselves simultaneously into each other.

By a *half rotation H* about a point M is to be understood a rotation through the angle π, i.e., a rotation which when performed again results in the identity. If A, B, C are three points such that A goes into C by a half rotation about B, and C goes into A by this rotation then B will be called the *center of the segment AC.*

If C is a point inside or outside the true circle drawn about A through B the segment AC will be said to be *smaller or greater* than the segment AB, respectively. In order to define the concept of "smaller" and "greater" analogously for arbitrary segments or circles perform motions under which the end points of the segments or the centers of the circles respectively go into the same point.

§ 24. A true segment AC has at most *one* midpoint. If there existed two midpoints for AC then denoting the half rotations about these midpoints by H_1 and H_2 the product $H_1H_2^{-1}$ would represent a motion which would leave each of the points A and C fixed, and thus by Section 22, it would be concluded, on denoting the identity symbolically by 1, that

$$H_1H_2^{-1} = 1, \quad \text{i.e.} \quad H_1 = H_2.$$

The centers thus coincide. In particular, the following result is deduced:

If two segments are congruent then their halves are also congruent.

§ 25. For the further developments the following lemma is required:

Let the points $A_1, A_2, A_3, \ldots$ converge to the point A and the points $M_1, M_2, M_3, \ldots$ to the point M. If the point A_i is carried into B_i by a half rotation about the point M_i then the points $B_1, B_2, B_3, \ldots$ converge, and indeed, to the point B which is generated from A by a half rotation about M.

It is possible to find a Jordan region in which lie the set of points $B_1, B_2, B_3, \ldots$. One convinces oneself of this by the same argument that was applied to the set of points $Z_1, Z_2, Z_3, \ldots$ in Section 21.

Denote now by B^* a limit point of the points $B_1, B_2, B_3, \ldots$. By Axiom III there must exist a motion which carries the points A, M, B^* into the points B^*, M, A, respectively, i.e., B^* is generated from A by a half rotation about M. Since, however, B is also generated from A by a half rotation about M it follows that $B^* = B$ and the proof is thus complete.

§ 26. *Let M be the midpoint of some segment AB. It will be shown that every segment AC which is smaller than AB also has a midpoint N.*

To do this draw any continuous curve γ from A to M and determine for every point M' of this curve the point B' such that M' becomes the midpoint of AB'. Then, as can be deduced from the lemma proved in Section 25, the locus of the points B' is a continuous curve γ'. The curve γ' terminates in the point A provided the point M' approaches A on the curve γ. If this is not the case assume that $M_1, M_2, M_3, \ldots$ is an infinite sequence of points on γ which converges to A, and that $B_1, B_2, B_3, \ldots$ are the corresponding points on γ'. If $B_1, B_2, B_3, \ldots$ have a limit point A^* that is distinct from A then it can be inferred that there exists a motion which leaves some arbitrarily close points to A arbitrarily close to A and at the same time brings the point A arbitrarily close to A^*. Then by Axiom III, A could remain fixed and at the same time be carried into A^* by some rotation. This, however, is impossible.

Since by hypothesis AC is smaller than AB the true circle drawn about A through C must meet the continuous curve γ' connecting A with B at some point B'. The point M' corresponding to this point on the curve γ is the midpoint of the true segment AB', and since $AC \equiv AB'$, the desired midpoint N, of AC, is found from M' by a suitable rotation about A.

Since by a half rotation about its midpoint N the segment AC goes into CA the following is deduced from the theorem proved above:

The segment AC is always congruent to the segment CA, provided that the segment AC is smaller than the segment AB assumed at the beginning of Section 26.

It can be seen at the same time that if the points $C_1, C_2, C_3, \ldots$ converge to the point A then the midpoints $N_1, N_2, N_3, \ldots$ of the segments $AC_1, AC_2, AC_3, \ldots$ converge to A.

§ 27. For the further developments some theorems on tangent true circles are necessary and the first thing that comes to mind is the *construction of two congruent circles which touch on the outside at one and only one point.*

To do this choose a circle χ' so small that no segment congruent to the one assumed in Section 26 lies in it. The theorem in Section 11 shows that this is possible since the points A and B can otherwise be moved arbitrarily close to M. Then let χ be a circle lying inside χ' about the same center as χ'. Take any two points on the circle χ and draw about them congruent circles α and β so small that any two points on χ which lie inside α can never

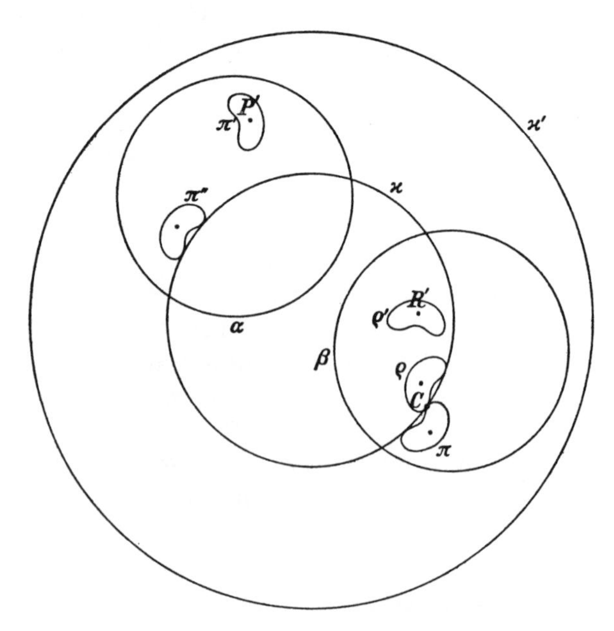

be separated by any two points on χ which lie inside β in the sense of the ordering of the points on χ. Furthermore, let the circles α, β be chosen so small that they lie entirely inside the circle χ'. Then take a point P' that lies inside α and outside χ and a point R' that lies inside β and inside χ and draw about P' and R' congruent circles π' and ρ' so small that π' lies entirely inside α and outside χ and further than ρ' lies entirely inside β and inside χ. Then perform a rotation about the center of α so that the circle π' goes into a circle π'' to which the circle χ is tangent from the outside. The points of contact form a set. Let it be denoted by S. Then perform a rotation about the center of β so that the circle ρ' goes into a circle ρ to which the circle χ is tangent from the inside. These points of contact form a set. Let it be denoted by T.

Since by the choice of the circles α, β no two points of the set S are separated by a pair of points of the set T on χ, then by a rotation of the plane about the center of the circle it is possible to cover one of the outermost points of S on χ with one of the outermost points of T on χ in such a way that the remaining points of S go into points all of which are distinct from the points of T. By this rotation the circle π'' comes in contact with the circle ρ in such a way that the point C at which the contact occurs is the only point of contact. Let the circle π'' in its new position be denoted by π and the centers of π and of ρ by P and R, respectively.

It will be shown now that the point of contact C must be the midpoint between the two centers, P, R. Indeed, in view of the choice of χ' the segment PR must be smaller than the fixed segment AB and therefore by Section 26 has a midpoint. Let it be denoted by C^*. Then by a half rotation about C^* each of the two circles π, ρ goes into each other. Since the point C is common to both circles π, ρ under such a half rotation it must also go into one of the common points of the circles π, ρ. Consequently it must remain invariant under this half rotation and thus must coincide with the point C^* about which the rotation is performed.

For the theorem proved above the following results can be seen at the same time:

The circle ρ which is tangent to the circle π from the outside at the point C is generated from π by a rotation about C on π. Besides ρ there exists no other circle that is congruent to π and tangent to it at the point C and only at this point from the outside.

§ 28. Furthermore, the following theorem is valid:

If any circle ι is surrounded and tangent to the circle π then the contact is only at one point.

For the proof assume that Q, Q' are two distinct points of contact of the circles ι and π. Make a half rotation about Q'. Under it π goes into a circle π' which is tangent to π only at the point Q' and ι goes into a circle ι' which lies inside π' and thus entirely outside π. Both circles π, π' are tangent only at Q'. Performing now the rotation about the center

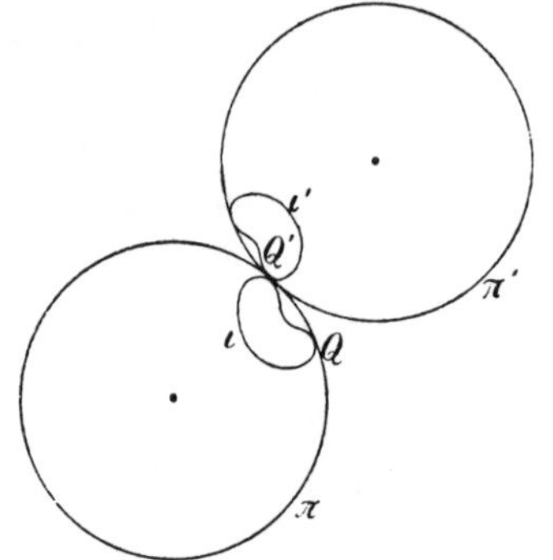

of the circle π by which Q goes into Q' a circle ι'' is generated from ι, which lies entirely inside π, and therefore also outside ι' to which it is tangent only at Q'. There are thus two circles ι, ι'' both of which are tangent to the congruent circle ι' from the outside at Q' and touch it only at this point. This contradicts the theorem in Section 27.

The results found in Sections 27 and 28 remain valid if instead of π, ρ smaller circles are taken.

§ 29. Let P be the center of the circle π constructed in Section 27 and Q a point on it. Furthermore, let O be any point. Then by resorting to the remark at the end of Section 26 and the theorem in Section 20 it is possible, as in Section 27, to determine a point E so close to O that inside the circle ι drawn about the midpoint M of the segment OE constructed through O and E, there exists no segment that is congruent to PQ. The same also holds for every point E' and the corresponding circle ι' if E' is situated closer to the point O than E.[1]

The following theorem then holds:

The circle ι (or ι') drawn about the midpoint M (or M') of OE (or OE') is completely surrounded by the circle about O through E (or E') and is tangent to it only at E (or E').

For the proof first construct the circle ω about O which surrounds the circle ι and at the same time is tangent to it. This circle ω is necessarily smaller than the circle π. For otherwise the circle congruent

[1] Choose a circle α about O in which lies no segment congruent to PQ. Denote by E a boundary point of this circle about O. Each of its interior and boundary points determines with O a segment whose midpoint M' lies in α. The circle about M' through O is congruent to the one about O and through M'. Consequently it contains no segment that is congruent to PQ.

to π drawn about O would lie in the interior of the circle ι, and thus there could exist a segment inside ι that is congruent to PQ, which is impossible. By the theorem proved in Section 28, the circle ω can have only one point of contact with ι. Let it be E_1. Now if E_1 is distinct from E perform a rotation about M by which E_1 approaches O. Under this rotation O moves then to a point E_2 of the circle ι which could be distinct from E_1. Since the segment OE_1 is congruent to the segment E_2O and thus also to OE_2 then E_2 would also have to be a point of the circle ω. This would contradict the fact that E_1 is the only common point of the circles ω and ι, i.e., the circle ω passes through E and the assertion is thus proved.

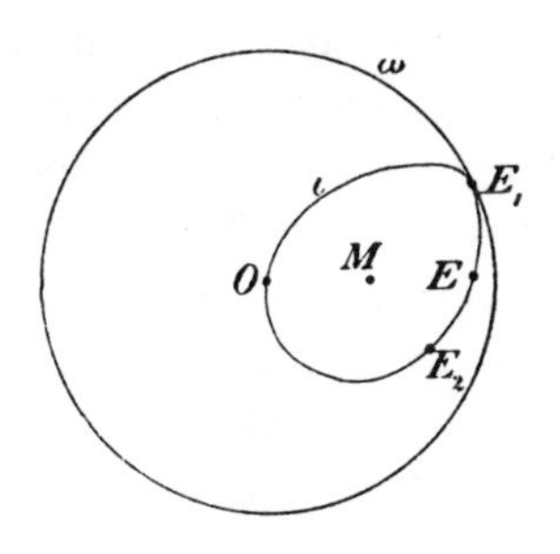

§ 30. In the following developments, begin with the segment OE constructed in Section 29, and assign the values 0 and 1 respectively to the points O and E. Construct then the midpoint of OE and assign to it the value 1/2. Then assign to the midpoints of the segments (0, 1/2) and (1/2, 1) the values 1/4 and 3/4, respectively. Then assign to the midpoints of the segments (0, 1/4), (1/4, 1/2), (1/2, 3/4), (3/4, 1) the values 1/8, 3/8, 5/8, 7/8, respectively, etc. Furthermore perform a half rotation of the whole segment (0, 1) about the point 0 and assign in general the numerical value $-a$ to the point that is generated from the point that corresponds to the number a. Then perform a half rotation about the point 1 and assign in general the numerical value $2 - a$ to the point that is generated from the point that corresponds to the number a. Consider such half rotations performed alternately about O and E and the newly generated points numbered correspondingly until every number a is assigned a definite point, whenever a is a rational number whose denominator is a power of 2.

§ 31. It is easy to realize the following rule from this correspondence:

By a half rotation about the point that corresponds to a number a every point x goes into the point $2a - x$. Hence if a half rotation is performed about the point $O = 0$ and then one about the point a then every point x is transformed into the point $x + 2a$.

§ 32. To order among themselves the points, to which numbers correspond, and to compare the segments which they bound among themselves, the theorem on tangent circles stated in Section 29 will be used in the following way:

The circle about the point 0 through the point 1/2 surrounds completely the circle about 1/4 through 1/2. Since it surrounds completely the circles about 1/8 through 2/8 = 1/4, and about 3/8 through 4/8 = 1/2, the last in turn surrounds the circles about 1/16 through 2/16 = 1/8, about 3/16 through 4/16 = 1/4, about 5/16 through 6/16 = 3/8, about 7/16 through 8/16 = 1/2, etc., it can be seen that the segment (0, 1/2) is greater than all segments (0, *a*) provided *a* is a positive rational number whose denominator is a power of 2 and whose value is less than 1/2.

Furthermore, the circle about 0 through 1/4 surrounds the one about 1/8 through 2/8 = 1/4. The second surrounded circle on its part surrounds the circles about 1/16 through 2/16 and about 3/16 through 4/16. These in turn surround the smaller circles about 1/32, 3/32, 5/32, 7/32, etc. Hence it can be seen that the segment (0, 1/4) is greater than all segments (0, *a*) provided *a* is a positive rational number whose denominator is a power of 2 and whose value is less than 1/4.

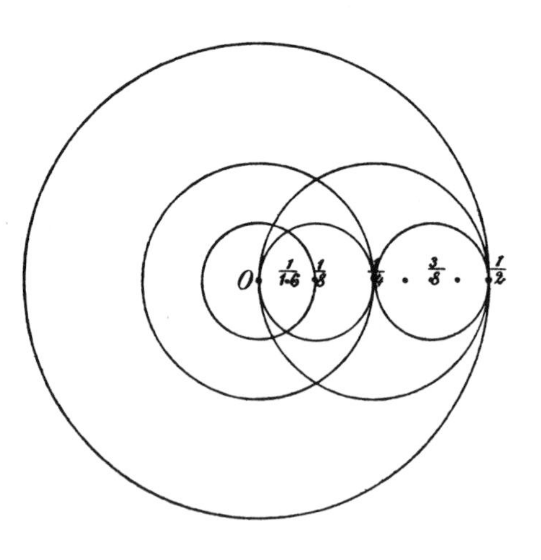

Consider next the circle about 0 through 1/8. It surrounds the circle about 1/16 through 2/16 = 1/8, and it in turn surrounds the smaller circles about 1/32 through 2/32, etc. Hence it is seen that the segment (0, 1/8) is greater than the segment (0, *a*) provided *a* is a positive rational number whose denominator is a power of 2 and whose value is less than 1/8. Continuing this line of reasoning the following general result is obtained:

If a is a rational number whose denominator is a power of 2 *and whose value is less than* $\frac{1}{2^m}$ *then the segment* (0, *a*) *is always smaller than the segment* $\left(0, \frac{1}{2^m}\right)$.

§ 33. The stage is set now to prove the following lemmas in succession:

The points which correspond to the numbers 1/2, 1/4, 1/8, 1/16, . . . *converge to the point* 0.

For otherwise, since the segments (0, 1/2), (0, 1/4), (0, 1/8), (0, 1/16), . . . are monotonically decreasing, the points 1/2, 1/4, 1/8, 1/16, . . . would have their limit point on a true circle χ about the point 0. Let then $\frac{1}{2^{n_1}}\ \frac{1}{2^{n_2}}\ \frac{1}{2^{n_3}}, \ldots$ be a sequence of points which converges to a point K on χ, and let the points

$$\frac{1}{2^{n_1+1}}, \quad \frac{1}{2^{n_2+1}}, \quad \frac{1}{2^{n_3+1}}, \ldots$$

converge to a limit point K^*. By the theorem in Section 25 it would follow that K^* could be the midpoint of the segment OK. By the result found at the end of Section 27 this would contradict the fact that K^* also lay on the circle χ.

§34. *Let* $a_1, a_2, a_3, \ldots$ *be positive rational numbers whose denominators are powers of* 2. *If the infinite sequence of numbers* $a_1, a_2, a_3, \ldots$ *converges to* 0 *then the sequence of points corresponding to these numbers also converges to* 0.

For the proof choose the integral exponents $n_1, n_2, n_3, \ldots$ such that

$$a_1 < \frac{1}{2^{n_1}}, \quad a_2 < \frac{1}{2^{n_2}}, \quad a_3 < \frac{1}{2^{n_3}}, \ldots$$

and such that the sequence $\frac{1}{2^{n_1}}, \frac{1}{2^{n_2}}, \frac{1}{2^{n_3}}, \ldots$ converges to 0. Since by the theorem in Section 32 the point a_i lies inside the circle 0 through $\frac{1}{2^{n_i}}$ and by the lemma proved in Section 33 the circles about 0 through $\frac{1}{2^{n_1}}, \frac{1}{2^{n_2}}, \frac{1}{2^{n_3}}, \ldots$ converge to 0, the assertion to be proved follows immediately.

§35. Finally the following theorem holds:

Let $a_1, a_2, a_3, \ldots$ *be an infinite sequence of rational numbers whose denominators are powers of* 2 *and which converges to some real number a. Then the corresponding points also converge to some definite point.*

For the proof, assume to the contrary that there exist two distinct limit points V' and V'' for the points $a_1, a_2, a_3, \ldots$. Let the points $a_1', a_2', a_3', \ldots$ converge to V' and $a_1'', a_2'', a_3'', \ldots$ to V''. By the remarks in Section 31 there exists for every point a_k a motion consisting of two half rotations which carries the point a_i' into the point $a_i' - a_k$ and at the same time carries the point a_i'' into the point $a_i'' - a_k$, and since the numbers $a_i' - a_k$ as well as the numbers $a_i'' - a_k$ with increasing indices also approach 0 arbitrarily close, it can be seen then by the theorem in Section 34 that there exist motions which take an arbitrarily close point to V' and at the same time also an arbitrarily close point to V'' arbitrarily close to the point 0. This, by a frequently applied argument based on Axiom III, is impossible.

§ 36. By assigning the number a to the point to which the points $a_1, a_2, a_3, \ldots$ converge then every real number is made to correspond to a definite point of the plane. The set of all these points will be called a *true line* so that *a true line is to be understood as the set of points which is generated from the points O, E by repeatedly taking midpoints, performing half rotations and adjoining all limit points of all resulting points. All sets of points resulting from motions of these true lines will also be called true lines. Every true line decomposes into two rays at every one of its points.*

§ 37. Using the lemma in Section 25 it is easy to see that under a half rotation about any point a of a true line a point x goes into the point $2a - x$. By performing two half rotations about the points 0 and a the point x goes into $x + 2a$.

It is easy to deduce from the theorem in Section 35 that even if a_1, $a_2, a_3, \ldots$ are any numbers converging to a then the corresponding points $a_1, a_2, a_3, \ldots$ always converge to the corresponding point a, *i.e., the true line is a continuous curve.*

§ 38. Suppose there existed two numbers a and b which represented the same point P of the plane on the true line. The point $\frac{a+b}{2}$ is the midpoint of the segment (a, b) and would have to coincide with the point P. The same would have to be true of the midpoints of the segments $\left(a, \frac{a+b}{2}\right)$ and $\left(\frac{a+b}{2}, b\right)$, i.e., be true of the points $\frac{3a+b}{4}$ and $\frac{a+3b}{4}$. On repeatedly taking the midpoints it is seen that all points $\frac{A_n a + B_n b}{2^n}$, where A_n, B_n are positive integers whose sum is 2^n, could be identical to P. It would thus follow by Section 37 that all real numbers lying between a and b could correspond to the same point P on the line. This contradiction shows that *the true line has no double points.* It can also be seen that *the true line cannot double back on itself.*

§ 39. *Two lines have at most one point in common.*

Indeed if they had the two points A and B in common and if to these points corresponded the numbers a and b on one line and the numbers a', b' on the other then by Section 24 the midpoints $\frac{a+b}{2}$ and $\frac{a'+b'}{2}$ could also coincide with each other. On repeatedly taking the midpoints as in Section 38 it can be concluded in a similar fashion that all points between a and b or a' and b' lie on both lines and thus these lines are identical.

§ 40. *The true line intersects every circle about any of its points, say the one about the point* 0.

Indeed, by the contrary assumption there are only two possible cases. There either exists a circle χ about the point 0 which the true line g meets but does not meet the circles about 0 surrounding χ, or there exists a circle χ which g does not meet but meets all circles about 0 lying inside χ.

Since by the nature of its construction the line g can always be extended from each of its points, and as is shown in Section 38 it can have no double point, then in the **first** case there could exist a circle about 0 inside χ which it met at two points AB on the same side of 0, where B is to be taken on the extension of g behind A and sufficiently close to A inside χ. Were a rotation performed about the point 0 by which A goes into B the line g would go into another line which besides intersecting g in 0 would also intersect it in B. By the theorem proved in Section 39 this would be impossible.

In the **second** case let K be a point of the circle χ to which the line g passes arbitrarily close. Draw a true circle π^* about K that is smaller than χ and which meets g, say at M. Then draw a circle π about M that is greater than π^* and smaller than χ. The circle π, since it is greater than π^*, contains the point K in its interior, and since it is smaller than χ the assumption made together with what has been proved above shows that the line g passing through M lies continuously inside π, extends from there in one or the other direction, leaves the circle π through a point on π and does return to the circle π. However, since the line g is supposed to approach the point K lying inside π arbitrarily close it must contain the point K itself. This is a contradiction of the present assumption.

Since the set of all circles about a point covers the entire plane without gaps it also follows from the foregoing that *two points of this plane geometry can always be connected by a true line.*

§ 41. It is only necessary to show now that *the congruence axioms in this plane geometry are valid.*

To do this choose a definite true circle χ and introduce for its points the parametric representation by the angle ω according to Section 18. Then if ω assumes the values of 0 through 2π the true circle will be oriented in a definite sense. With the introduction of this parameter every other circle congruent to χ will also obtain a definite orientation, namely the one that results from covering the center of the circle χ with the center of the given circle by two successively applied rotations in accordance with Section 22. Since in view of the definition of a rotation at the beginning of this discussion it is impossible to cover the original circle χ with itself in the opposite orientation, there exists indeed a definite orientation for every circle.

Take now two rays emanating from the same point M which together do not form a true line, construct about M a circle congruent to χ and fix the part of the circle delineated by the ray which corresponds to the parametric interval that is numerically less than π. The set orientation leads then along the fixed arc from one of the two rays to the other. Call the first ray the right-hand side and the second ray the left-hand side of the angle between the two rays and the parametric interval ($< \pi$) itself the value of the angle. From the definition of a motion the first congruence theorem for two triangles follows then in the following form:

If the congruences

$$AB \equiv A'B', \quad AC \equiv A'C', \quad \measuredangle\, BAC \equiv \measuredangle\, B'A'C'$$

hold for two triangles ABC *and* $A'B'C'$, *and furthermore, if* AB, $A'B'$ *and* AC, $A'C'$ *are the corresponding right- and left-hand sides of the angles* $\measuredangle BAC$ *and* $\measuredangle B'A'C'$ *respectively, then the congruences*

$$\measuredangle\, ABC \equiv \measuredangle\, A'B'C' \quad \text{and} \quad \measuredangle\, ACB \equiv \measuredangle\, A'C'B',$$
$$BC \equiv B'C'$$

always hold.

§ 42. Having defined and derived the properties of the true line in Sections 30-40 there are two cases to be distinguished.

Assume first, that through one point there exits only *one* line which does not intersect a given line (axiom of parallels). Then for this plane all plane axioms which I formulated in the main part of this book (Chapter I) are valid except that the congruence axiom III, 5 formulated previously in Section 41 is to be taken there in the narrower form. Even with this narrower form of this congruence axiom, the Euclidean plane geometry necessarily follows (cf. Appendix II, p. 114 as well as Chapter I, pp. 27-28).

Secondly, assume that through every point A there exist two rays which together do not form one and the same line and which do not intersect a line g whereas every ray lying in the angle space formed by them and emanating from A does intersect the line g. A thus lies outside of g.

With the aid of continuity it easily follows then conversely, that to any two rays emanating from a point A which together do not form one and the same line, there always corresponds a line g which does not intersect these two rays, but which meets every other ray that emanates from A and lies in the angle space of the two given rays. In this case the Bolyai-Lobachevskian plane geometry follows even if Axiom III, 5 is taken in the previously formulated narrower form, as can be shown with the aid of my "end" arithmetic.[1]

In conclusion, I shall point out the characteristic difference that exists between the present development of geometry and the one I attempted to give in the main part of this book. There the arrangement of the axioms is such that continuity is required **last** among the axioms so that then the question as to what extent the well-known theorems and arguments of elementary geometry which are independent of continuity arises in the foreground in a natural way. In the present investigation, however, continuity is required **first** among the axioms by the definition of the plane and a motion so that here the most important task has been rather to determine the least number of conditions from which to obtain by the most extensive use of continuity the elementary figures of geometry (circle and line) and their properties necessary for the construction of geometry. Indeed, the

[1] Compare my article "A New Development of Bolyai-Lobachevskian Geometry," Appendix III of this book. The argument presented there should be suitably modified for this purpose so that continuity is used and an application of the theorem on the equality of the base angles in an isosceles triangle is avoided. In order to obtain the theorems for the addition of ends (pp. 143-145), consider addition as the limiting case of a rotation of the plane as the point of rotation recedes along a line to infinity.

present investigation has shown that to this end the conditions stated in Axioms I-III are sufficient.

Göttingen, May 10, 1902

A survey of the development of the researches which followed up Riemann's and Helmholtz's investigations of the foundations of geometry with an accompanying bibliography is given by H. Freudenthal in the introduction to his article "Neuere Fassungen des Reimann-Helmholtz-Lieschen Raumproblems," *Math. Zeitschrift,* Vol. 63, 1955/1956, as well as in the article "Im Umkreis der sogenannten Raumprobleme," *Essays on the Foundations of Mathematics*, Jerusalem, 1961.

APPENDIX V

SURFACES OF CONSTANT GAUSSIAN CURVATURE

Surfaces of Constant Negative Curvature

According to Beltrami[1] a surface of constant negative curvature realizes a region of a Lobachevskian (non-Euclidean) plane if the geodesics of the surface are taken as the lines of the Lobachevskian plane and the actual lengths and angles on the surface are taken as the lengths and angles in the Lobachevskian plane. Among the surfaces of constant negative curvature investigated thus far *none* has been found which can be extended continuously everywhere by a continuous variation of its tangent plane in the neighborhood of every point. Rather, the known surfaces of constant negative curvature have singular lines beyond which extensions with continuously varying tangent planes are impossible. For this reason no one has succeeded in realizing the *entire* Lobachevskian plane with the hitherto known surfaces of constant negative curvature and the question of fundamental interest arises *whether the entire Lobachevskian plane can be represented at all by Beltrami's method through an analytic*[2] *surface of constant negative curvature.*

In order to answer this question a surface of constant negative curvature -1, which in every finite region is everywhere regular and has no singularities, will be assumed at the outset. It will then be shown that this assumption leads to a contradiction. A surface such as the one to be assumed is completely characterized by the following statement:

Every limit point of points of the surface that lies in a finite region is also a point of the surface.

[1] *Giornale di Matematiche*, Vol. 6 (1868).

[2] For the sake of simpler expressions I am assuming analyticity for the surface to be considered although the arguments and the result obtained (cf. p. 197) remain valid if **P** (x, y) in equation (1) is a sufficiently differentiable nonanalytic function of x, y. That there actually exist nonanalytic regular surfaces of constant negative curvature in the sense of surface theory (which according to the subsequently proved theorem can neither be extended continuously everywhere with a continuously varying tangent plane) has been proved at my urging by G. Lütkemeyer in his Inaugural Dissertation "Über den analytischen Charakter der Integrale von partiellen Differentialgleichungen," (Göttingen, 1902).

If O is any point of the surface then it is always possible to set up orthogonal coordinate axes x, y, z in such a way that O is the origin of the coordinate system and the equation of the surface in the neighborhood of O is

$$z = ax^2 + by^2 + \mathbf{P}(x, y), \tag{1}$$

where the constants a, b satisfy the relation

$$4ab = -1$$

and the power series $\mathbf{P}(x, y)$ contains only powers of x, y of third and higher orders. Clearly the z-axis is then along the normal to the surface and the x-axis and the y-axis give the directions by which the principal curvatures of the surfaces are determined.

The equation

$$ax^2 + by^2 = 0$$

determines the two principal tangents* of the surface through the point O in the xy-plane. These, therefore, are always separated and give the directions of the two asymptotic curves of the surface through the arbitrary point O. Each of these is one of a simple family of asymptotic curves which covers the entire neighborhood of the point O on the surface in a regular manner and without gaps. Taking u and v as sufficiently small values the following construction can be performed: Lay off along one of the two asymptotic curves passing through O a length that is equal to the parametric value of u; draw through the end point thus obtained the other possible asymptotic line and lay off along it a length that is equal to the parametric value of v. The point obtained now is a point of the surface that is uniquely determined by the parametric values u, v. If the orthogonal coordinates x, y, z of the surface are regarded accordingly as functions of u, v by setting

$$x = x(u, v), \quad y = y(u, v), \quad z = z(u, v),$$

then for sufficiently small values of u, v these are regular analytic functions of u, v.

*These are the asymptotes of the Dupin indicatrix or of the surface whose order of contact with the given surface is at least 2. (Translator's note)

The familiar theory of surfaces of constant curvature -1 provides the following additional results:

If φ is the angle between the two asymptotic curves through the point (u, v) then the three quantities of the first fundamental form of the surface assume the values

$$e \equiv \left(\frac{\partial x}{\partial u}\right)^2 + \left(\frac{\partial y}{\partial u}\right)^2 + \left(\frac{\partial z}{\partial u}\right)^2 = 1,$$

$$f \equiv \frac{\partial x}{\partial u}\frac{\partial x}{\partial v} + \frac{\partial y}{\partial u}\frac{\partial y}{\partial v} + \frac{\partial z}{\partial u}\frac{\partial z}{\partial v} = \cos\varphi,$$

$$g \equiv \left(\frac{\partial x}{\partial v}\right)^2 + \left(\frac{\partial y}{\partial v}\right)^2 + \left(\frac{\partial z}{\partial v}\right)^2 = 1,$$

and thus the square of the derivative with respect to the parameter t of the arc length of any curve on the surface takes the form

$$\left(\frac{ds}{dt}\right)^2 = \left(\frac{du}{dt}\right)^2 + 2\cos\varphi\,\frac{du}{dt}\frac{dv}{dt} + \left(\frac{dv}{dt}\right)^2. \tag{2}$$

As a function of u, v the angle φ satisfies the partial differential equation

$$\frac{\partial^2\varphi}{\partial u \partial v} = \sin\varphi.\text{[1]} \tag{3}$$

On dropping the one-to-one correspondence requirement between pairs of values, u, v and points of the plane, the performed construction can be extended to any values of u, v. To be sure, the u-curve passing through O can even be closed. But in any case, by the assumption made for the surface (p. 192), arbitrarily large lengths can be laid off along it on both sides of O. Thus to every value of u corresponds a point on the asymptotic curve.

Consider now the other asymptotic curve passing through every such point P. Along this one take as the parameter v the length determined from P (in one direction). Again it is possible to lay off arbitrarily large distances along the asymptotic curve on both sides of P.

[1] On the basis of this formula, I first demonstrated the impossibility of the existence of a surface on constant negative curvature that is free of singularities (*Trans. Am. Math. Soc.*, Vol. 2, 1901). Later E. Holmgren gave a more extensive analytic proof of this proposition that is also based on Formula (3) *Comptes Rendus* (Paris, 1902). The adaptation of Holmgren's proof that follows in the text, fits in with the presentation which W. Blaschke gave in his *Vorlesungen über Differentialgeometrie*, I, Section 80 (1921). Along with my original proof note also the approach by L. Bieberbach, *Acta Mathematica*, Vol. 48.

In this way, to every pair of values u, v, a point of the surface corresponds in a single valued manner, but generally not in a single valued and invertible manner. Geometrically speaking, a mapping of the entire Euclidean (u, v)-plane into some covering surface of the given surface or a region thereof is thus obtained.

The problem now is to show that every u-curve on the surface is an asymptotic curve and that the parameter u represents its arc length.

For the curve $v = 0$ this is already known. Furthermore, by the arc element representation (2) this is also valid for arcs of v-curves which lie in the neighborhood of a point $(u, 0)$.

For the general proof it is sufficient to show the following:

If a is a positive number and b is any real number then the image of every segment

$$-a \leqq u \leqq +a, \quad v = b$$

on the surface is an arc of an asymptotic curve or a union of arcs along such a curve and u represents its length.

Next, this proposition holds for $b = 0$. Furthermore, the following can be shown:

1. If the proposition holds for $b = b_0$ then it also holds for every b that differs sufficiently little from b_0.

2. If the proposition holds for $b_1 < b < b_2$ then it also holds for $b = b_1$ and for $b = b_2$.

This can be done by continuity arguments and an application of the Heine-Borel covering theorem.

The proof is thus given for all b.

Now if $\varphi = \varphi(u, v)$ (as on p. 193 f.) is the angle between the two asymptotic curves passing through the point (u, v) of the surface, measured from the positive u-direction to the positive v-direction, then $\varphi(u, v)$ is a continuous function defined for all values of (u, v) with continuous partial derivatives which satisfy the **differential equation** (3).

By a suitable choice of the positive u-direction and v-direction it can be deduced that the inequalities

$$0 < \varphi < \pi \quad \text{and} \quad \frac{\partial \varphi}{\partial u} \geqq 0$$

hold **at the point** $u = v = 0$.

Since φ nowhere equals 0 or π then in view of the continuity of the function $\varphi(u, v)$,

$$0 < \varphi(u, v) < \pi,$$

and thus

$$\sin \varphi > 0$$

for all values of u, v.

However, a function $\varphi(u, v)$ with these properties cannot exist.

Since from the differential equation

$$\frac{\partial^2 \varphi}{\partial u \partial v} = \sin \varphi,$$

it follows that

$$\frac{\partial^2 \varphi}{\partial u \partial v} > 0;$$

it also follows that $\frac{\partial \varphi}{\partial u}$ increases with increasing v.

In particular

$$\frac{\partial \varphi}{\partial u}(0, 1) > \frac{\partial \varphi}{\partial u}(0, 0) \geqq 0$$

and hence it is possible to determine a positive quantity a such that for $0 \leqq u \leqq 3a$:

$$\frac{\partial \varphi}{\partial u}(u, 1) > 0.$$

Let m denote the positive minimum of

$$\frac{\partial \varphi}{\partial u}(u, 1) \qquad \text{for} \qquad 0 \leqq u \leqq 3a.$$

Then for $v \geqq 1$:

$$\left.\begin{aligned} \varphi(a, v) - \varphi(0, v) &= \frac{\partial \varphi}{\partial u}(\vartheta a, v) \cdot a \\ &\geqq \frac{\partial \varphi}{\partial u}(\vartheta a, 1) \cdot a \geqq m \cdot a \end{aligned}\right\} \quad (0 < \vartheta < 1)$$

and also

$$\varphi(3a, v) - \varphi(2a, v) \geqq m \cdot a,$$

so that

$$\varphi(a, v) \geqq \varphi(0, v) + m \cdot a > m \cdot a$$

and

$$\varphi(2a, v) \leqq \varphi(3a, v) - m \cdot a < \pi - m \cdot a.$$

Furthermore, for $0 \leqq u \leqq 3a, \quad v \geqq 1$:

$$\frac{\partial \varphi}{\partial u}(u, v) \geqq \frac{\partial \varphi}{\partial u}(u, 1) > 0,$$

so that $\varphi(u, v)$ increases monotonically with u. Hence for

$$a \leqq u \leqq 2a, \quad v \geqq 1:$$

$$0 < m \cdot a < \varphi(a, v) \leqq \varphi(u, v) \leqq \varphi(2a, v) < \pi - m \cdot a,$$

so that

$$\sin \varphi(u, v) > \sin(m \cdot a) = M,$$

where $M > 0$ and is independent of u, v.

Therefore the value of the double integral

$$\iint \sin \varphi(u, v)\, du dv,$$

taken along the rectangle with the vertices

$$(a, 1), \quad (2a, 1), \quad (2a, V), \quad (a, V), \quad (V > 1)$$

is greater than

$$M \cdot a \cdot (V - 1),$$

and so by a suitable choice of V is greater than π.

On the other hand, from the differential equation (3) one obtains

$$\iint \sin \varphi\, du\, dv = \int_a^{2a} \int_1^V \frac{\partial^2 \varphi}{\partial u \partial v}\, du dv$$

$$= (\varphi(2a, V) - \varphi(a, V)) - (\varphi(2a, 1) - \varphi(a, 1)) < \pi,$$

since

$$\varphi(2a, V) - \varphi(a, V) < \varphi(2a, V) < \pi$$

and

$$\varphi(2a, 1) - \varphi(a, 1) > 0.$$

A contradiction is thus reached and the basic assumption must be dropped, i.e., it is seen *that there exists no singularity free and everywhere regular analytic surface of constant negative curvature. In particular, the question posed at the outset, whether the entire Lobachevskian plane can be realized by a regular analytic surface in space by Beltrami's method, is thus to be answered in the negative.*

Surfaces of Constant Positive Curvature[1]

The beginning of this investigation started out with the quest for a surface of constant negative curvature which is everywhere regular and analytic in every finite region and the result was that such a surface does not exist. The same problem for positive constant curvature will be treated now with a corresponding method. Clearly, the sphere is a closed surface of positive constant curvature, free of singularities, and according to the proof devised by Liebmann[2] at my urging, there exists no other closed surface with the same property. This result will be formulated as a theorem that is valid for any portion of a surface of positive curvature[3] free of singularities to read as follows:

Let a finite simply or multiply connected region, free of singularities, be delineated on a surface of positive constant curvature + 1. *Suppose that the two principal radii of curvature are constructed at every interior, as well as at every boundary point of the region. Then the maximum of the larger, and hence the minimum of the smaller, of the two radii, of principal curvature are not attained at any point that lies in the interior of the region; for let the surface be a portion of the sphere of unit radius.*

[1] The problem of realizing the non-Euclidean plane elliptic geometry with the points of an everywhere continuous curved surface has been investigated at my urging by W. Boy in "Über die Curvatura integra und die Topologie geschlossener Flächen," Inaugural Dissertation (Göttingen, 1901), and *Math. Ann.*, Vol. 57, 1903. In this work W. Boy constructed a topologically very interesting finite surface closed on one side, which, with the exception of a closed double curve with a triple point in which the sheets of the surface intersect, has no singularities, and has the same connectivity as the non-Euclidean elliptic plane.

[2] *Göttinger Nachrichten* (1899), p. 44. Cf. also the interesting works of the same author in *Math. Ann.*, Vol. 53 and Vol. 54.

[3] G. Lütkemeyer, in the Inaugural Dissertation noted on p. 191, and E. Holmgren, in *Math. Ann.*, Vol. 57, demonstrated the analytic character of surfaces of constant positive curvature.

For the proof, first note that in view of the hypothesis, the product of the radii of principal curvature is equal to 1 everywhere, and hence the larger of the two radii of principal curvature must be greater than or equal to 1. Hence the maximum of the larger of the radii of principal curvature can evidently be equal to 1 only if both radii of principal curvature are equal to 1 at every point of the surface region. In this special case every point of the surface region is an umbilic and one concludes easily in the familiar way, that the surface region must be a portion of a sphere of unit radius.

Now let the maximum of the two radii of principal curvature of the surface be greater than 1. Then assume contrary to the assertion that there exists in the **interior** of the surface region a point O at which that maximum is attained. Since this point cannot be an umbilic and is otherwise a regular point of the surface the neighborhood of this point is covered simply and without gaps by each of the two families of lines of curvature of the surface. Taking these lines as curvilinear coordinates and the point O as the origin then by the familiar theory of surfaces of constant positive curvature the following hold:[1]

Let r_1 denote the larger of the two radii of principal curvature for the point (u, v) in the neighborhood of the origin $O = (0,0)$. In this neighborhood $r_1 > 1$. Setting

$$\varrho = \tfrac{1}{2}\log\frac{r_1 + 1}{r_1 - 1};$$

the positive real quantity ρ as a function of u, v satisfies the partial differential equation

$$\frac{\partial^2 \varrho}{\partial u^2} + \frac{\partial^2 \varrho}{\partial v^2} = \frac{e^{-2\varrho} - e^{2\varrho}}{4}. \tag{4}$$

Since ρ necessarily increases with decreasing r_1 then as a function of u, v it must attain a minimum at $u = 0$, $v = 0$, and hence, the expansion of ρ in powers of the variables u, v must have the form

$$\varrho = a + \alpha u^2 + 2\beta uv + \gamma v^2 + \cdots,$$

where a, α, β, γ are constants and thus the quadratic form

$$\alpha u^2 + 2\beta uv + \gamma v^2$$

can never assume negative values for real u, v. From the last condition

[1] Darboux, "Leçons sur la théorie générale des surfaces," Vol. 3, No. 776; Bianchi, "Lezioni di geometria differenziale," Section 264.

the inequalities

$$(5) \qquad \alpha \geqq 0 \quad \text{und} \quad \gamma \geqq 0$$

necessarily follow for the constants α and γ.

On the other hand let the expansion of ρ be substituted in the differential equation (4). For $u = 0, v = 0$ one obtains

$$2(\alpha + \gamma) = \frac{e^{-2a} - e^{2a}}{4}.$$

Since the constant a represents the value of ρ at the point $O = (0, 0)$ and thus is positive, the expression on the right-hand side is at any rate less than 0. The last equation leads then to the inequality

$$\alpha + \gamma < 0,$$

which contradicts equation (5). Hence, the original assumption, that the maximum is attained at a point in the interior of the surface region, is untenable, and the theorem formulated above is thus seen to be true.

Hence, as already noted above, follows immediately the theorem that a *closed surface free of singularities and of constant positive curvature* 1 *must be a sphere of unit radius.* This result expresses at the same time the fact that the sphere as a whole cannot be bent without producing a singularity somewhere on its surface.

Finally the above analysis leads to the following results for a surface that is not closed. If a portion is cut out from the surface of the sphere and bent arbitrarily then the maximum of all radii of principal curvature always occur on the boundary of the portion of the surface.

Göttingen, 1900.

SUPPLEMENTS

by P. Bernays

SUPPLEMENT I

1. Remarks on Sections 3-4

In Chapter I at the end of Section 3 (p. 5) it is remarked that the fact that a line a cannot intersect all three sides of a triangle ABC is a provable theorem. It behooves one to deduce the proof by means of Theorem 4. It can be carried out as follows: If the line a met the segments BC, CA, AB at the points D, E, F then these points would be distinct. By Theorem 4 one of these points would lie between the other two.

If, say, D lay between E and F, then an application of Axiom II, 4 to the triangle AEF and the line BC would show that this line would have to pass through a point of the segment AE or AF. In both cases a contradiction of Axiom II, 3 or Axiom I, 2 would result.

Van der Waerden observed that Pasch's Axiom II, 4 can be replaced by the following space order axiom:

II, 4*. *If A, B, C are three noncollinear points and α is a plane in which none of the points A, B, C lies and if the plane α passes through a point of the segment AB then it also passes through a point of the segment AC or BC.*

With this substitution not only II, 4 becomes a provable theorem but the incidence axiom I, 7 as well. For this see Van der Waerden's article "De logische Grondslagen der Euklidische Meetkunde" which appeared in *Zeitschrift Christian Huygens* Vols. 13-14 (1934-1936), Section 3.

For the proof of the assertion of I, 7 with Axiom II, 4* let the following be observed: Van der Waerden uses in his proof the axiom that every plane contains three noncollinear points. This axiom, which also appeared in Hilbert's previous editions of *Foundations of Geometry*, is replaced by the weaker requirements in Axioms I, 3 and I, 4 (cf. Section 3) that there exists three noncollinear points to begin with and that in every plane there exists at least one point. With this choice of axioms it is possible to prove I, 7 from II, 4* by first proving from II, 4* the following restricted theorem: "If a plane β which contains three noncollinear points has a point in common with a plane α then α and β

have another point in common." One shows then with the aid of this theorem and the incidence axiom I, 1-6 and I, 8 that every plane contains three noncollinear points whereby the restrictive condition for the plane β can be removed.

The fact that the assertion of Axiom I, 7 can be proved from II, 4*, which expresses the fact that space is at most three dimensional, is analogous to the fact that Axiom II, 4, if the requirement in it that line a lie in the plane ABC is dropped, restricts the dimensionality to two.

For Theorem 9 in Section 4 (p. 9) there is a detailed proof by G. Feigl in his article "Über die elementaren Anordnungssätze der Geometrie," *Jahresbericht der Deutschen Mathematische - Vereinigung,* Vol. 33 (1924), Section 4.

---

2. Remarks on Section 13

The set of axioms for the real numbers given in Section 13 is essentially taken from Hilbert's article "Über den Zahlbegriff," *Jahresbericht der Deutschen Mathematische - Vereinigung,* Vol. 8 (1900), in which the requirements listed in Section 13 as theorems are stated as axioms. The following remarks are cited here from that article:

1. The existence of the number 0 (Theorem 3, p. 44) is a consequence of Theorems 1 and 2 and the associative law of addition.
2. The existence of the number 1 (Theorem 6, p. 44) is a consequence of Theorems 4 and 5 and the associative law of multiplication.
3. The commutative law of addition (Theorem 8, p. 45) is a consequence of Theorems 1-6, the associative law of addition and the two distributive laws, namely,

$$(a+b)(1+1) = (a+b)1 + (a+b)1 = a+b+a+b,$$
$$= a(1+1) + b(1+1) = a+a+b+b;$$

whence

$$a+b+a+b = a+a+b+b,$$

and thus by Theorem 2

$$b + a = a + b.$$

That the commutative law of multiplication (Theorem 12, p. 45) can be deduced from Theorems 1-11, 13-16 and 17 (Archimedes' Theorem), but not without the use of 17, is shown in Sections 32 and 33.

SUPPLEMENT II

A Simplified Development of the Theory of Proportion

The development of the theory of proportion without the use of the Archimedean Axiom, i.e., on the basis of Axioms I, 1-3, II-IV, as given in Chapter III, Sections 14-16, can be simplified.

The notation introduced at the beginning of Section 15 on p. 55 for segments, segment equality, segment sums, as well as the fact, as pointed out there, that segment sums follow the associative and the commutative laws, will be used here.

Define now the ratio $a : b$ of the segments a, b as the angle (that is uniquely determined in the sense of congruence) which lies opposite the leg a in a right triangle whose legs are a, b. Segment ratios will be said to be equal if their defining angles are congruent and in this sense a "proportion" shall be denoted as $a : b = c :$ d. Hence follows without further comment that every segment ratio is equal to itself and that two segment ratios which are equal to a third are equal to each other, and also that:

if $\quad a = c$ and $b = d$, then $a : b = c : d.$

By the theorem on the sum of the angles of triangles it also follows that

if $\quad a : b = c : d$, then $b : a = d : c.$

Furthermore, the following theorem can be deduced by means of Axioms III, 5 and IV (see the figure). If

$a : b = c : d$, then

$a : b = (a + c) : (b + d)$,

and by exterior angle theorem, if

$a : b = a : c$, then $b = c.$

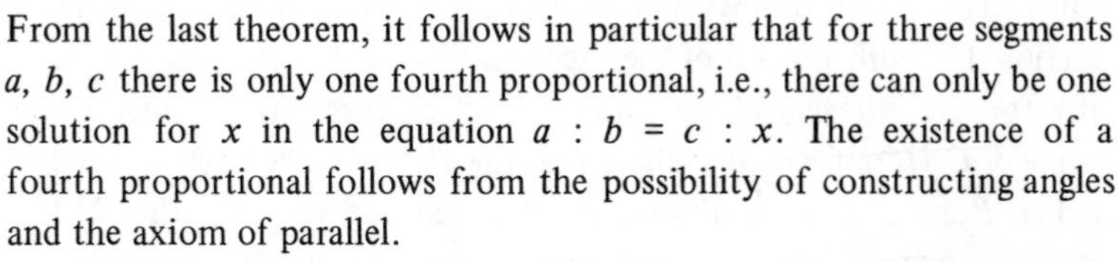

From the last theorem, it follows in particular that for three segments a, b, c there is only one fourth proportional, i.e., there can only be one solution for x in the equation $a : b = c : x$. The existence of a fourth proportional follows from the possibility of constructing angles and the axiom of parallel.

The theorem on the interchangeability of the inner terms in a proportion, i.e., the assertion that if $a : b = c : d$ then $a : c = b : d$, follows by considering a right triangle whose legs are a, b and another one whose legs are c, d and which are so positioned that the leg

c of the second triangle lies at the vertex of the right angle of the first triangle along the extension of the leg b, and that d lies similarly along the extension of a (see the figure). By the assumed proportion the end points of both hypotenuses lie on a circle. This can be deduced from the theorem on the equality of peripheral angles subtending equal arcs or from its converse. The proportion $a : c = b : d$ thus also follows by this theorem by considering the triangles whose legs are a, c and b, d.[1]

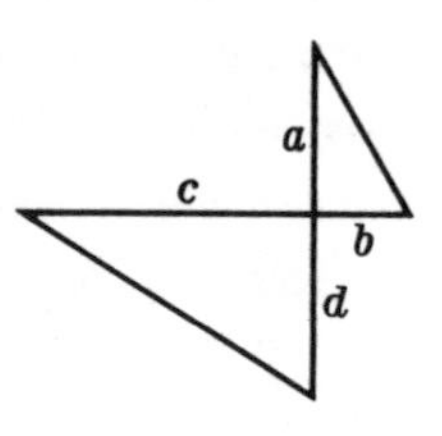

As a consequence of the interchangeability of the inner terms of a proportion one obtains in particular the possibility of combining proportions, namely,

If $a : b = a' : b'$ and $b : c = b' : c'$, then $a : c = a' : c'$.

In other words, from the assumed proportions one obtains, by interchanging the inner terms, $a : a' = b : b' = c : c'$ and hence $a : c = a' : c'$.

By using the existence of the fourth proportional one also obtains the following second combination rule:

If $a : b = b' : a'$ and $b : c = c' : b'$, then $a : c = c' : a'$.

In other words, if u is the fourth proportional to a, b, c', so that $a : b = c' : u$, then it follows by the above assumption that $c' : u = b' : a'$; that is, $c' : b' = u : a'$. Furthermore, $u : a' = b : c$ and by combining the proportions $c' : u = a : b$ and $u : a' = b : c$ (by the preceding rule) it follows that $c' : a' = a : c$.

It remains now to prove the fundamental theorem of proportion. If two parallels delineate the segments a, a' on one side of an angle and b, b' on the other side of the angle, then the proportion $a : a' = b : b'$ holds. The proof follows (similarly to the proof of Theorem 41 of Section 16) with the aid of the theorem that the bisectors of the angles of a triangle intersect at the same point. This theorem is applied to the triangles OAB and $OA'B'$, where O is the vertex of the angle of interest and A, B and A', B' are the points at which one and the other of two parallels, respectively, intersect the sides of the angle. Furthermore, $OA = a$, $OB = b$, $OA' = a'$, $OB' = b'$ (see the figure). Let S be the point of

[1] The form of this proof is taken from the text *Fragen der Elementargeometrie* by F. Enriques.

intersection of the bisectors of the angle OAB and S' that of the triangle $OA'B'$. Then S is at the same distance r from the lines OA and OB. S' is also at the same distance r' from these lines. It will be shown that $a : a' = r : r'$. In the same way it is found then that $b : b' = r : r'$ and the assertion thus follows.

The proportion $a : a' = r : r'$ is obtained by considering the triangles OAS and $OA'S'$ for which $\measuredangle SOA = \measuredangle S'OA'$ and $\measuredangle SAO = \measuredangle S'A'O$. Moreover, both of these angles (as halves of one angle) are acute. The feet D and D' of the perpendiculars from S and S' to the line OA thus fall in the interior of the segments OA and OA'. OA is partitioned into the sections u, v and OA' into the sections u', v'. Then $r : u = \measuredangle SOA = \measuredangle S'OA' = r' : u'$, $r : v = \measuredangle SAO = \measuredangle S'A'O = r' : v'$. Hence $r : r' = u : u' = v : v' = (u + v) : (u' + v') = a : a'$, so that indeed $a : a' = r : r'$.

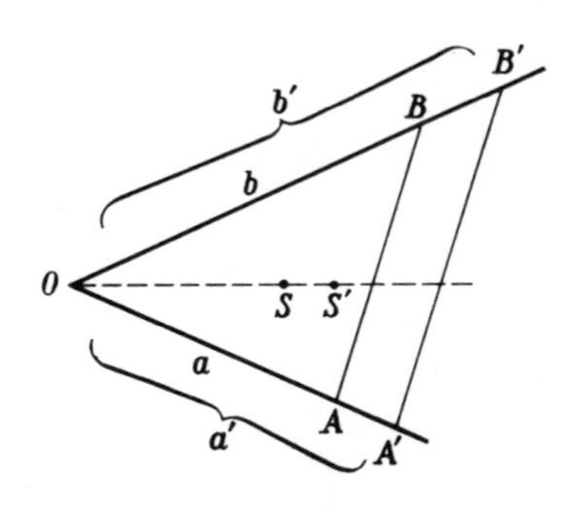

With the use of the fundamental theorem of the theory of proportion the proposition referred to by Hilbert as Pascal's Theorem, in particular, can be reduced to the second combination rule of proportions mentioned above. That theorem (Theorem 40 in Section 14) states that if A, B, C are three points on one line and A', B', C' are three points on another line, which intersect, and all these points are distinct from the point of intersection of the two lines then AB' is parallel to BA' provided BC' is parallel to CB' and CA' is parallel to AC'. Denoting the segments OA, OB, OC by a, b, c, respectively, and the segments OA', OB', OC' by a', b', c', respectively, then according to the fundamental theorem of the theory of proportion, the hypothesis is equivalent to the two proportions

$$b : c = c' : b' \quad (1) \qquad\qquad c : a = a' : c' \quad (2)$$

and the asserted consequence is equivalent to the proportion

$$a : b = b' : a'.$$

This one, however, can be obtained from the preceding two by the second combination rule.

Having thus developed the theory of proportion without the use of segment multiplication the latter can now be introduced with the establishment of a unit segment e. In other words, the product of two segments a, b (with respect to the unit segment e) is defined as the fourth proportional to e, a, b.

As far as the rules of operation of the segment multiplication thus defined are concerned, the commutative law follows from the interchangeability of the inner terms in a proportion. The associative law states that if $e : a = b : u, e : b = c : v$ and $e : u = c : w$ then $e : a = v : w$. This can be obtained as follows: From the second assumed proportion one obtains $b : e = v : c$. This together with the third assumed proportion yields, by a combination, $b : u = v : w$. Hence, by the first assumed proportion, $e : a = v : w$. The distributive law states that if $e : a = b : u$, $e : a = c : v$ then $e : a = (b + c) : (u + v)$. For the proof it is sufficient to show that if $b : u = c : v$ then $b : u = (b + c) : (u + v)$. This, however, holds by a theorem mentioned at the outset.

It is possible to associate now a plane analytic geometry, as in Section 17, with the segment arithmetic thus introduced.

REMARK. Instead of defining segment multiplication by establishing a unit segment for the purpose of obtaining an analytic geometry, it is possible to introduce a ratio arithmetic. Addition and multiplication of ratios are then defined by the following rules:

$$(a : c) + (b : c) = (a + b) : c,$$ [1]

$$(a : b) \cdot (b : c) = a : c;$$

equal ratios added to equal ratios yield equal ratios, and equal ratios multiplied by equal ratios yield equal ratios.

That these rules can be satisfied and that the remaining rules of addition and multiplication are also valid, follows from the previously (pp. 203-04) established properties of proportions, in particular from the existence of the fourth proportional. Without further comments one thus obtains the inverse of multiplication, namely, division by $a : b$ is multiplication by $b : a$, and the ratio $a : a \ (= b : b)$ plays the role of the multiplicative identity.

[1] The inadequacy of defining the sum of two segment ratios as the sum of the ratios of the defining angles, i.e., to define it by the geometric addition of angles, stems from the fact that the defining angle of a segment ratio is always acute, whereas the sum of two acute angles need not be an acute angle.

SUPPLEMENT III

Remarks on the Theory of Plane Areas

Some clarifying and sharpening comments will be given to the definitions of equidecomposability and equicomplementability and to the subsequent theorems in Section 18 (pp. 60-61). The term "polygon" will always be used here in the sense of a "simple polygon."

First let it be noted that for a more precise formulation of the definition of the "decomposition" of a polygon P into two polygons P_1, P_2 (cf. p. 60) it is necessary to incorporate in it the recursive definition of decomposition into several polygons.

The consequences which can immediately be related in Section 18 to equidecomposability and to equicomplementability are generally valid provided only that the concepts of "addition" and "removal" are understood in the restricted sense. If it is desired not to be restricted in the formulation it behooves one to introduce the general concept of a "polygonal" alongside that of a polygon. As a starting point for this it is possible to use the **theorem** that every simple polygon can be decomposed into triangles.

In order to prove this theorem by induction on the number of sides of the polygon it is sufficient to show that every polygon P whose number of sides is greater than three can be decomposed into two polygons with a smaller number of sides.

Let $A_1, A_2, \ldots, A_n$ be the successive vertices of a polygon P. Consider the ray emanating from A_1 into the interior of the polygon. Let it first meet the polygon again at B. (The ray can always be chosen so that B does no lie on one of the lines A_1A_2, A_1A_n.) In case B is one of the vertices of P then the segment A_1B yields immediately a decomposition of P into two polygons with a smaller number of sides. Otherwise B lies on a polygon side A_iA_{i+1}, where $2 \leqq i \leqq n-1$ and the equality sign holds at most once. If $2 < i < n-1$, then the segment A_1B yields again a decomposition of the desired type. There remain then the cases $i = n-1$ and $i = 2$, which can be treated in the same way. Suppose that $i = 2$. Then the polygon P is decomposed by the segment A_1B into the triangle A_1A_2B and the n-gon $A_1BA_3 \ldots, A_n$. Let the latter be denoted by P'. The segment A_1A_3 decomposes it now into a triangle and an $(n-1)$ a-gon, or P' lies in the interior of the triangle A_1BA_3 (which in any case contains interior points of P') or at least one vertex A_j with $j > 3$ lies on the side A_1A_3 of that triangle. In the first case the desired decomposition is obtained

Otherwise, from among the rays emanating from A_1 and passing through one of the vertices A_j take the one whose point of intersection with the line BA_3 is the closest to B, and on this ray take the vertex of P' lying closest to A_1. Let A_k be that vertex. If $k \neq n$, then the segment A_1A_k decomposes the polygon P' into two polygons with a smaller number of sides. For $k = n$, the segment BA_k yields such a decomposition.[1]

The decomposition of a polygon into triangles produces no **overlaps**, that is, no two triangles have an interior point in common. By additional partitions, as necessary, it is also possible to satisfy the sharper **triangulation property** whereby every pair of triangles is either disjoint, has only one vertex in common, or has one side and otherwise no other points in common.

A collection (Gesamtheit) of a finite number of triangles in the plane which has the triangulation property[2] need not triangulate a polygon. The figures which can be triangulated by such collections of triangles are of a more general nature. Let them be called **polygonals.**

In order to obtain the figure of a polygonal from a collection of triangles which triangulates it, first it is necessary to omit the sides of those triangles which are common to two triangles. The remaining broken line then possibly forms a simple polygon in which two or more adjacent sides of a triangle which lie on the same line can be formed into a side of a polygon. However, the broken line decomposes (by the indicated formation method) into several simple polygons. Let the sides of these polygons also be called the sides of the polygonal. A partition of the points of the plane outside these polygons into "exterior" and "interior" points of the polygonal can be made in the following way: The finite collection of triangles can be enclosed by a polygon, or even by a triangle. The points outside the polygon which enclose the polygonal are exterior points of the polygonal. Otherwise the distinction between interior and exterior points is determined by the fact that every crossing of a side of the polygonal leads either from exterior to interior points or from interior to exterior points. Accordingly one convinces oneself that the exterior points are all those which lie outside

[1] A different argument for the proof of the decomposability of every simple polygon into triangles can be found in the article by Van der Waerden quoted in Supplement I, 1. See Section 5, Theorem 24.

[2] In topology such a collection is called a finite plane triangle complex.

the collection of the triangles and that the interior points are those which either lie inside one of these triangles or on a side of these triangles but not on a side of the polygonal.

In this way every finite collection of **nonoverlapping** triangles determines a polygonal, as from it, by a suitable partition, it is possible to obtain a collection of triangles which has the triangulation property and the resulting polygonal is independent of the various possible choices of the partition. In other words, for every two partitions it is always possible to determine a third, that is a partition of each of the other two. A collection of triangles that results from a triangulation of a polygonal by a partition yields again the same polygonal since the sides of the triangles that are added in the partition occur twice in the narrower triangulation.

From two nonoverlapping polygonals P, Q it is possible to obtain a new polygonal by combining the collection of triangles of a triangulation of P with those of a triangulation of Q whereby a nonoverlapping collection of triangles is obtained which determines a polygonal. The polygonal obtained in this manner is independent of the particular triangulations of P and Q, as can be deduced from the preceding remarks about the partitions. It is thus uniquely determined by P and Q. It is said to be obtained from a **combination** of P with Q, or from P by the **adjunction** of Q, and is denoted by "$P + Q$."

This process of combination can be extended to several nonoverlapping polygonals in pairs; and this combination is associative and commutative, since the union of nonoverlapping collections of triangles in pairs has these properties.

Let now P and Q be any two polygonals given by the triangulating collection of triangles $\Delta_1, \ldots, \Delta_k$ and $\Delta'_1, \ldots, \Delta'_l$. Then for every triangle Δ_i it is possible to determine a triangulating partition $\Delta_{i1}, \ldots, \Delta_{ir_i}$ such that no triangle Δ_{ij} is intersected by a side of a triangle Δ'_h and moreover that every triangle Δ_{ij} is either entirely in a triangle Δ'_h or has no common inner point with it. In this manner one obtains a triangulation of P in which every triangle is either entirely in one of the triangles $\Delta'_1, \ldots, \Delta'_l$ or has with each of these no inner point in common. The triangles of the first kind yield together a triangulation of a polygonal T_1, and those of the second kind a triangulation of a polygonal T_2 and $P = T_1 + T_2$. If the triangles of the triangulation of T_1 in Δ'_h are now separated from it the remaining figure can be triangulated and can thus be seen to be a polygonal R_h. A combination of the nonoverlapping polygonals $R_1, \ldots, R_l$ in pairs yields a polygonal T_3 and $Q = T_1 + T_3$. The polygonals T_1, T_2, T_3 are thus nonoverlapping in pairs. Their combination $T_1 + T_2 + T_3$ will be defined as the **union** of P with Q. T_1 will be called the **intersection** of P with Q.

For a more precise treatment of this analysis the following situation should be used: If a vertex or a point of a side of a triangle lies in the interior of a polygonal, then an interior point of the triangle also lies in the interior of the polygonal. Hence it follows that if in two triangulated polygonals P, Q a triangle of the triangulation of P has no interior point in common with a triangle of the triangulation of Q, then P and Q are nonoverlapping.

As a result of the last discussions note the following: For any two polygonals P,Q there always exists a union V and an intersection D with the following properties: V and D are polygonals. It is possible to determine polygonals P' and Q' such that P', Q', D are nonoverlapping in pairs and

$$P = D + P', \quad Q = D + Q', \quad V = D + P' + Q',$$

and also

$$V = P + Q', \quad V \doteq Q + P'.$$

Equidecomposability and equicomplementability of polygonals will now be defined.

DEFINITION. Two polygonals P, Q will be said to be equidecomposable if a triangulation $\Delta_1, \ldots, \Delta_n$ for P and a triangulation $\Delta'_1, \ldots, \Delta'_n$ for Q can be determined such that Δ_i and Δ'_i are congruent triangles. Two polygonals P, Q will be said to be equicomplementable if it is possible to adjoin to P a polygonal P' and to Q a polygonal Q' such that P' is equidecomposable with Q' and $P + P'$ is equidecomposable with $Q + Q'$.

To justify these definitions it must be shown that they are consistent with the previous definitions of equidecomposability and equicomplementability of polygons (Section 18, p. 60).

As far as equidecomposability is concerned it is seen immediately that the new definition for polygons has the same meaning as the former. Now the **transitivity of equidecomposability**, or the theorem that two polygonals which are equidecomposable with a third are equidecomposable with each other, can be proved quite analogously to the previous one for polygons (cf. the proof on p. 61). Without further comments one also obtains the theorem of the **additivity of equidecomposability**. If a polygonal P is composed of the polygonals P_1 and P_2 and similarly a polygonal Q is composed of the polygonals Q_1 and Q_2 and if P_1 is equidecomposable with Q_1, and P_2 is equidecomposable with Q_2, then P is equidecomposable with Q.

In order to show now that the definition of equicomplementability of polygonals is equivalent to the previous definition for polygons P, Q the following **lemma** will be established: Let P be a polygon and K a polygonal. Then it is possible to adjoin to P a polygon H that is equidecomposable with K so that $P + H$ is a polygon. Briefly, P can be extended to a polygon by adjoining to it a polygon H that is equidecomposable with K.

For the proof, consider first the case in which a side a of the polygon P lies on a line g such that all vertices of the polygon, with the exception of the endpoints of a, lie on one and the same side of g. The polygonal K is triangulated by the triangles $\Delta_1, \ldots, \Delta_k$. Every triangle is equidecomposable now with a rectangle. This can be seen in the following way: Let ABC be the triangle and let the angles at A and B be acute. If p is the connecting line of the midpoints of CA and CB and if F and G are the feet of the perpendiculars from A and B to p then the rectangle $ABGF$ is equidecomposable with the triangle ABC, as can be seen by dropping a perpendicular from C to FG. (Cf. the figure for the proof of Theorem 39 on p. 38.)

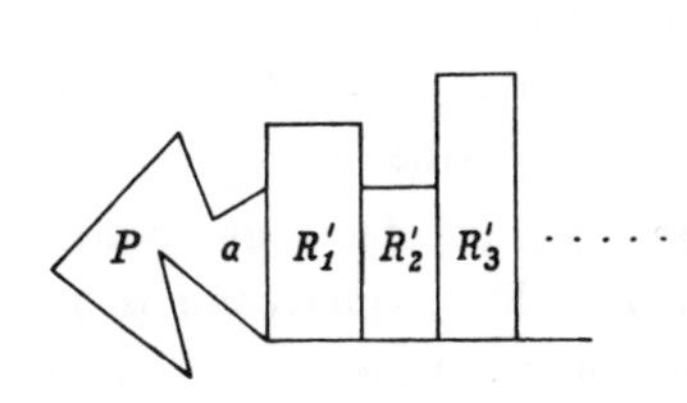

A rectangle R_i is thus equidecomposable with each of the triangles Δ_i ($i = 1, \ldots, k$) and it is possible to attach to the polygon P the rectangles $R'_1, R'_2, \ldots, R'_k$ successively so that R'_i is congruent to R_i and $R'_l + \ldots + R'_k$ as well as $P + R'_l + \ldots + R'_k$ are polygons. The polygon $R'_l + \ldots + R'_k$ is then equidecomposable with the polygonal K, whereby the assertion in this case is proved.

The general case can be reduced to this special case in the following way: Through a vertex of the polygon P draw a line q on which lies no other vertex P and which is parallel to no line connecting two vertices of P. Through all vertices of P draw the parallels to q and intersect all these parallels $q_1, \ldots, q_r$ with a line c. Among the points of intersection two are the outermost. Let q_i be one of the lines $q_1, \ldots,$ q_r which passes through one of the outermost points of intersection. Then only one vertex E_i of P lies on q_i and all remaining vertices of P lie on one and the same side of q_i. It is possible now to adjoin to the polygon a triangle Δ with a vertex E_i so that $P + \Delta$ is a polygon again and a side of Δ lies on the line q_i whereas the vertex of Δ opposite that side lies on a side of P terminating in E_i. At the same time it is possible to choose Δ so small that a decomposition of K in the form $K = \Delta' + K'$ is possible wherein Δ and Δ' are congruent. The conditions for the above special case are now satisfied for the polygon $P + \Delta$ and the polygonal K'. According to what has been proved for this case it is then possible to determine for $P + \Delta$ a polygon H that is equidecomposable with K' so that $P + \Delta + H$ is a polygon. Hence by the argument of the above proof $\Delta + H$ is also a polygon. Moreover, $\Delta + H$ is equidecomposable with $\Delta + K'$ and so with K, whereby the assertion is proved.

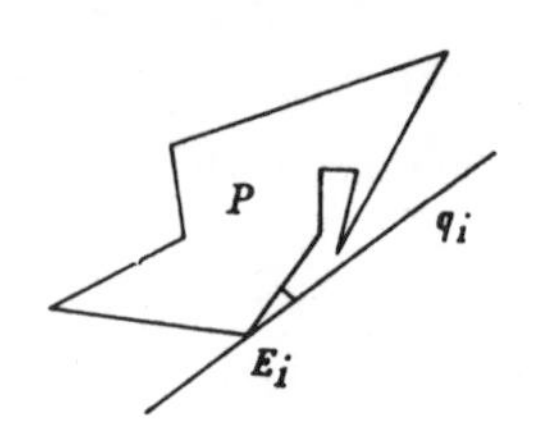

Let now P and Q be polygons that are equicomplementable in the sense of the definition of equicomplementability of polygonals. Then there exist equidecomposable polygonals P', Q' such that $P + P'$, $Q + Q'$ are equidecomposable polygonals again. According to the lemma it is possible to determine a polygon P'' that is equidecomposable with P' and a polygonal Q'' that is equidecomposable with Q' so that $P + P''$ and $Q + Q''$ are polygons. $P + P'$ is thus equidecomposable with $P + P''$ and $Q + Q'$ is equidecomposable with $Q + Q''$. By the transitivity of equidecomposability it follows now from these equidecomposabilities that P'' is equidecomposable with Q'' and $P + P''$ is equidecomposable with $Q + Q''$ But then the condition for equicomplementability in the sense of the definition of Section 18 is satisfied. Conversely, that two polygons which are equicomplementable according to this definition are also equicomplementable in the sense of the definition of equicomplementability of polygonals, is clear without further comments. The definition of equicomplementability of polygonals as applied to polygonals has thus the same meaning as the previous one.

At the same time this proof yields the following theorem: If P, Q are equicomplementable polygons then it is possible to adjoin to P a polygon P' and to Q a polygon Q' that is equidecomposable with P' such that $P + P'$ and $Q + Q'$ are equidecomposable polygons.[1]

By extending the concept of equicomplementability to polygonals it will be shown now that two polygons which are equicomplementable with a third are equicomplementable with each other.

Let the polygons P and Q be both equicomplementable with the polygon R. Then there exist polygonals P', Q', S, T with the property that P' is equidecomposable with S; Q' is equidecomposable with T; and that $P + P'$ and $R + S$ as well as $Q + Q'$ and $R + T$ are equidecomposable polygonals.

This shows at the same time that P and P' as well as Q and Q' are nonoverlapping, and that R and S as well as R and T, and hence R and the union of S with T, are also nonoverlapping. This union, as has been shown before, can be represented on one hand in the form $S + T'$ and on the other hand in the form $T + S'$ where T' and S' are some polygonals.

It is possible now to determine for the polygonal $P + P'$, which can be enclosed by a triangle, a polygonal P'' which is equidecomposable with T', lying entirely outside $P + P'$ and for $Q + Q'$ a polygonal Q'' which is equidecomposable with S', lying outside $Q + Q'$. By the additivity of equidecomposability there exists equidecomposability between $P' + P''$ and $S + T'$, between $Q' + Q''$ and $T + S'$, between $P + P' + P''$ and $R + S + T'$, and between $Q + Q' + Q''$ and $R + T + S'$. Furthermore, $S + T' = T + S'$. Hence follows that $P' + P''$ is equidecomposable with $Q' + Q''$ and that $P + P' + P''$ is equidecomposable with $Q + Q' + Q''$ so that P and Q are equidecomplementable.

Note that for this proof the lemma is not required, but rather the fact that for every polygonal, it is possible to determine another that is equidecomposable with it, lying entirely outside a given polygon, and that every polygonal can be enclosed with a polygon (triangle). Besides this, the transitivity and the additivity of equidecomposability of polygonals is also used.

With these means it is also possible to deduce the additivity of equicomplementability. Let P, Q, S, T be polygonals such that P, Q as well as S, T are nonoverlapping. Furthermore, let P be

[1] In previous editions, this condition for P and Q was taken as the property defining equicomplementability.

equicomplementable with S, and Q with T. It will be shown that $P+Q$ is also equicomplementable with $S + T$.

By hypothesis there exist polygonals P', Q', S', T' such that there are no overlaps between P and P', Q and Q', S and S', T and T', and that P' is equidecomposable with S', Q' with T', as well as $P + P'$ with $S + S'$, and $Q + Q'$ with $T + T'$.

It is possible now to determine a polygonal P^* that is equidecomposable with P' which lies outside the polygonal $P + Q$ and a polygonal Q^* that is equidecomposable with Q' which lies outside the polygonal $P + Q + P^*$. It is also possible to determine polygonals S^* and T^*, such that S' is equidecomposable with S^*, T' with T^*, and such that S^* lies outside $S + T$, and T^* outside $S + T + S^*$.

By the assumed equidecomposabilities and the transitivity and additivity of equidecomposability P^* is now equidecomposable with S^*, Q^* with T^*, $P + P^*$ with $S + S^*$, $Q + Q^*$ with $T + T^*$ and hence $P^* + Q^*$ is also equidecomposable with $S^* + T^*$ and $P + Q + P^* + Q^*$ with $S + T + S^* + T^*$. Hence, however, it follows that $P + Q$ is equicomplementable with $S + T$.

SUPPLEMENT IV

1. A Note on the Introduction of a Segment Arithmetic Based on Desargues' Theorem

In the development of a segment arithmetic independently of the congruence and continuity axioms, as it has been done in Sections 24-27 of Chapter V, no use was made of the order axioms II. Rather, the proofs are carried out with the aid of the incidence axioms I, 1-3, the axiom of parallels IV* (which is also an incidence axiom), and Desargues' Theorem in the form of Theorem 53 alone (p. 72). The axioms of order are first brought in, in Section 28, in order to show that once a segment arithmetic has been developed it is possible to define a quantitative relation of segments for which the order rules 13-16 of Section 13 are satisfied. If these rules are disregarded then instead of a Desarguesian number set, as it is defined at the end of Section 28, one obtains the concept of a *skew field*, i.e., a number set which has all the properties of a number field (rational domain) except the commutativity of multiplication (1-11 in Section 13).

Starting out with a skew field it is possible to obtain, as noted briefly at the beginning of Section 29 (cf. pp. 85-86), an analytic geometry of three dimensional space. For this geometry all Axioms I and the axiom of parallels IV* are satisfied. On the other hand, as noted in Section 22, it is possible to prove Desargues' Theorem from these axioms. One obtains thus, in correspondence to Theorem 56 (p. 88), the following theorem:

THEOREM 56*. *Let Axioms* I, 1-3 *and* IV* *be satisfied in some plane geometry. Then the validity of Desargues' Theorem is a necessary and sufficient condition that this geometry be embeddable in a space geometry in which Axioms* I *and* IV* *hold.*

2. Remarks on Section 37

In Chapter VII, Section 37, the assertion of Theorem 65 (p. 103) needs the following correction: The number of solutions to the construction problem, in case it can be performed by drawing lines and constructing segments under the given condition for the number n, can also be 2^{n+1}.

A simple example is the problem to determine for two points A, B a point C such that the triangle ABC has a right angle at A and the segments AB and AC are equal. Hence $n = 0$ and the number of solutions is 2.

This observation is due to Dono Kijne who analyzed the problem in its generality in his dissertation "Plane Construction Field Theory" (Utrecht, 1958).

SUPPLEMENT V

1. Equidecomposability in the Models of Appendix II

In Appendix II it is shown, among other things, that the development of the theory of area without the use of the Archimedean axiom is no longer possible by the method presented in Chapter IV (Sections 18-21) if the triangle congruence axiom III, 5 is replaced by the narrower congruence axiom III, 5* which is restricted to the case of correspondence of triangles of like orientation. This is also impossible even if assertions III, 6 and III, 7 for angle congruence (cf. p. 114) (which can be proved with the aid of III, 5) are taken as axioms as well as the requirement for the existence of the right angle. The exposition of this will be expanded whereby the result will be sharpened.

The discussions in Appendix II are related to two models of plane geometry which can be constructed by a common principle, each consisting in an analytic geometry in which the number set has all the properties of an ordered number field and which, however, is not the set of real numbers. In one case it is a non-Archimedean set with a generating parameter t which plays the role of an "infinitesimally small number" and in the other it is a countable subset of the real numbers. The definitions of incidence and order and also of parallelism are the usual ones. Only the definition of rotation has an anomaly by the fact that in rotating a vector through an angle α a positive numerical factor enters which depends on α. Figures (segments, angles, polygons) are said to be "congruent" if they can be carried into each other by parallel displacements and rotations.

It is shown that in these models, besides demonstrating the existence of the right angle, Axioms I, 1-3, II, II, 1-4, 5*, 6-7, IV are satisfied. In view of this, the theorems of the theory of congruence hold **insofar as corresponding figures of like orientation** (i.e., under the retention of "right" and "left") **are concerned.** Angles can be compared, the equality of right angles holds and the angle sum theorem in a triangle is valid.

Even some theorems which are usually proved by the use of congruences with different orientations, as well as the theorems that the perpendicular bisectors of a triangle intersect at one point, are also valid. These can be obtained from the following reflection theorem which follows directly from the analytic representation of a reflection. The combination of three reflections in three lines with the direction angles $\alpha_1, \alpha_2, \alpha_3$ passing through a common point S yields a reflection in the line with the direction angle $\alpha_1 - \alpha_2 + \alpha_3$ passing through S.

The theory of circles can also be derived with the aid of the aforementioned theorems, by defining a "circle over the segment AB" as the geometric locus of the points C for which ACB is a right angle. The equality of all inscribed angles over the same arc of the circumference can then also be shown, and one thus obtains all means by which to develop the theory of proportion, according to the method of Chapter III, or of Supplement II.

This method of showing the validity of the theory of proportion in the two considered models used in the previous editions of Hilbert, can indeed be replaced by a more direct one,[1] namely, by noting that in both models the following theorem of analytic geometry is valid: If the points P_1 and P_2 with the coordinates x_1, y_1 and x_2, y_2, respectively, lie on a ray emanating from the point P_0 with the coordinates x_0, y_0 then the ratio of the segments P_0P_1 and P_0P_2 is equal to the ratio of the abscissa differences $x_1 - x_0$ and $x_2 - x_0$, provided these are not equal to zero, and is equal to the ratio of the ordinate differences $y_1 - y_0$ and $y_2 - y_0$ provided these are not equal to zero.

The difference between these two geometries and the ordinary Euclidean geometry shows up in particular in the figure (cf. p. 125) constructed from two right triangles OQP and OQR which are the mirror images of each other and in which the angle QOP can be chosen so that the length of OR is smaller than that of OP as well as that of OQ. (In the case of the first non-Archimedean model the difference in lengths is only "infinitesimally small." However, in the second model it can be arbitrarily large.)

The fundamental principle of measuring length according to which the length of a segment is determined by its projections on the x-axis and the y-axis, i.e., Pythagoras' Theorem as a theorem on the lengths of segments in a right triangle, is thus not satisfied here. This motivated Hilbert to call this geometry a "non-Pythagorean geometry." On the other hand, however, Hilbert shows that Pythagoras' Theorem as a **theorem on equicomplementability** is valid in the constructed geometry, i.e., it is also true here that the square constructed on the hypotenuse of a right angle is equicomplementable with the squares on the legs.

[1] For this reason the discussion of the noted theorem on reflections and its corollaries has been eliminated from the Appendix of the Seventh Edition.

Indeed, Euclid's proof of this equicomplementability uses only congruences of corresponding figures of like orientation. Defining for this geometry, as already noted, congruence of polygons with the aid of congruent mappings (cf. pp. 116-17) i.e., by saying that polygons are congruent if they can be carried into each other by congruent mappings, and defining otherwise equidecomposability and equicomplementability as before, then Euclid's argument remains applicable. (Note in this case in particular that the proof of the transitivity of equicomplementability, which must be used by the method given in Supplement II, also remain valid in the narrower sense of the definition of congruence.)

For the given figure *OPQR* it follows in particular that the square constructed on *OP* is equicomplementable with the square constructed on *OR*, although the segment *OR* is shorter than *OP* so that the square constructed on *OP* is equicomplementable with a square embedded in it.

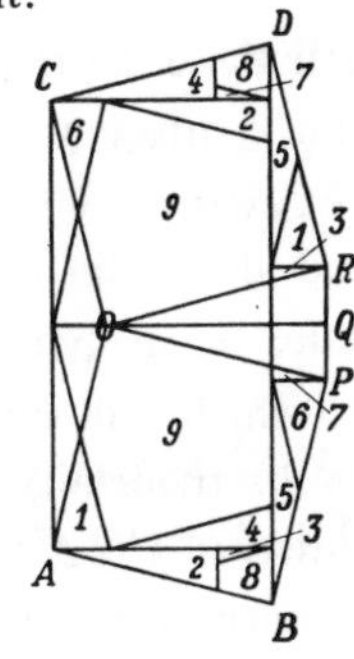

This corollary can even be sharpened now. For as it is well known not only equicomplementability can be proved for the square constructed on the hypotenuse of a right triangle with the squares on the legs but equidecomposability as well, whereby corresponding triangles and quadrangles of **like orientation** are even associated by parallel displacement.[1] Hence follows by the transitivity of equidecomposability that in the given figure *OPQR* the square over the segment *OP* is **equidecomposable** with the square over the segment *OR*.

Finally, this equidecomposability can also be seen directly from the accompanying figure, in which there is given for each of the squares *OABP* and *OCDR* a decomposition into nine polygons consisting of seven triangles, one quadrangle and one pentagon. From them, corresponding polygons of the two squares, denoted by the same numbers, can be obtained from each other by parallel displacement.

[1] For the accompanying figure, the title "The Chair of the Bride" has been handed down by the Hindus of the Ninth Century. **Alnairizi**, the Arabian commentator on Euclid, also established this equidecomposability.

If the shorter segment OR is constructed on OP from O to a point P' then it follows that the square $OP'B'A'$ constructed on OP', which is embedded in the square on OP, is equidecomposable with it.

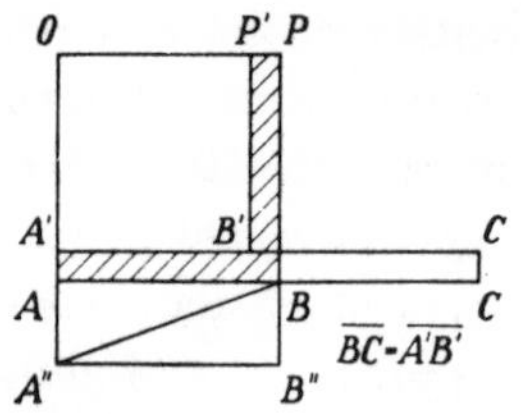

The following reasoning can also be applied to this case: The hexagon $AA'B'P'PB$ in which the lengths of the sides AB and BP are OP and $A'B'$, respectively, and the length of $B'P'$ is OR, is equidecomposable with the rectangle $ACC'A'$, where the point C is on the extension of AB from B and its distance from B is $A'B'$, and C' is the foot of the perpendicular from C to the line $A'B'$. This rectangle is equidecomposable in turn with the triangle $AA''B$, where A'' is a point on the extension of the segment OA from A. This can be shown with the usual elementary methods by using the fact that the length of the segment AC is less than twice the length of AB.

Denoting the square $OABP$ by T, the square $OA'B'P'$ by T', the hexagon $AA'B'P'PB$ by W, and the triangle $AA''B$ by Δ, then T is equidecomposable with T', Δ with W, and thus also $T + \Delta$ with $T' + W$, i.e., with T.

Extending now also the segment PB from B to B'' by the segment that is congruent to AA'' then the triangle $A''B''B$, which will be denoted by Δ', is congruent (with like orientation) to the triangle Δ. Hence $T + \Delta + \Delta'$ is equidecomposable with $T + \Delta$. However, $T + \Delta + \Delta'$ is the rectangle $OA''B''P$ and $T + \Delta$ is obtained from this rectangle by removing from it the triangle Δ'. A counter example to Theorem 52 has thus been found in this geometry.

Finally, by still starting out from the figure $OPQR$ it is possible to arrive at a configuration of two rectangles lying in each other and equidecomposable with each other in a way that is simpler than considering the squares on OP and OR. Rotating the rectangles $OQRS$, where S is the point on the perpendicular to OQ through O at a distance $\overline{QR}$ from O, through the angle QOP in the negative direction, one obtains a rectangle $OUQ'V$ in which U lies on the segment OP and Q' lies on OQ.

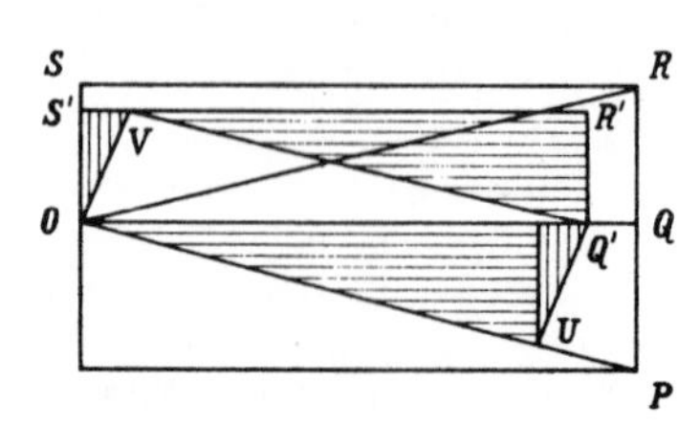

This rectangle in turn, as can be seen in a simple way, is

equidecomposable with the rectangle $OQ'R'S'$, where R' and S' lie on the parallels to OQ through V and S' lies on OS. This rectangle is thus equidecomposable with the rectangle $OQRS$ in which it is embedded.

2. Hilbert's Axiom of Embedment

In the previous editions, Appendix II contained a proof, in connection with the theory of area, that the adoption of an axiom of embedment that is equivalent to Theorem 52 makes it possible to deduce the original broader congruence axiom III, 5 from the narrower congruence axiom III, 5*. This argument will be given here with little insignificant changes adapted to the new edition.

It has been pointed out above (cf. p. 113) that congruence axiom III, 5, and thus in general the congruence of figures in the **broader** form, necessarily follows from the triangle congruence axiom in the **narrower** form III, 5* and from the preceding axioms together with III, 6 and III, 7, as soon as the validity of the theorem on the equality of the base angles of an isosceles triangle is assumed.[1] It appears to me noteworthy that this extension of the congruence axiom in the **narrower** form III, 5* can also be made in quite a different way, namely by means of a very intuitive requirement whose content essentially coincides with Theorem 52 (p. 69) proved by me in *Foundations of Geometry* and which on the other hand, as shown in Appendix II (cf. p. 127), is not a consequence of the congruence theorems in the **narrower** sense.

Let "equidecomposability" and "equicomplementability" for polygons be defined as in Section 18. Furthermore, let "combination" and "equidecomposability" of "polygonals" be defined as in Supplement III where, however, congruence is to be understood in the **narrower** sense. Accordingly, in what is to follow, the triangle congruence axiom, always in the **narrower** form III, 5*, and the preced-

[1] For a more precise exposition see the article by P. Bernays, "Bemerkungen zu den Grundlagen der Geometrie," *Courant Anniversary Volume* (1948), pp. 29-44. How the postulation of the base angle theorem, by including the Archimedean Axiom and the axiom of parallels, can be replaced in various ways by requirements which have no characteristic of a symmetry axiom, e.g., the requirement of the third congruence axiom for triangles of like orientation, is shown in A. Schmidt's Dissertation "Die Herleitung der Spiegelung aus der ebenen Bewegung," *Math. Ann.*, Vol. 109 (1934), pp. 538-71.

ing Axioms I, 1-3, II, III, 1-4, as well as the axiom of parallels IV, will be used.

"The requirement in question, that should serve as an extension, reads as follows:

AXIOM OF EMBEDMENT. *A polygon is never equidecomposable with a polygon whose boundary contains interior points but no exterior points of the first polygon, i.e., which is embedded in the first one.*

From this axiom the following theorem is deduced first:

A polygon is never **equicomplementable** with a polygon in which it is embedded."

Indeed, were a polygon P equicomplementable with a polygon Q lying in it then there could exist two equidecomposable polygonals P' and Q' such that $P + P'$ and $Q + Q'$ were equidecomposable polygons. P' and Q would then be nonoverlapping and in view of the additivity of equidecomposability $Q + P'$ and $Q + Q'$ would be equidecomposable, and by the transitivity of equidecomposability $P + P'$ would also be so with $Q + P'$. Since, on the other hand, Q lies inside P there exists a combination $P = Q + R$ where R is a polygonal. According to the lemma in Supplement III (cf. p. 211) there could be adjoined to $P + P'$ a polygon P'' that is equidecomposable with R and such that $P + P' + P''$ is a polygon. In view of the equidecomposability of $P + P'$ with $Q + P'$ and of P'' with R as well as the additivity of equidecomposability it would follow that $P + P' + P''$ were equidecomposable with $Q + P' + R$ and hence with $P + P'$. This, however, would contradict the axiom of embedment.

"Now the following theorems will be proved:

If the angles at *A* and *B* in a triangle *ABC* are equal then the sides opposite them are also always equal.

For the proof determine the points E and D on AB such that $AD = BC$ and $BE = AC$. By the first congruence axiom in the narrower form follows the congruence of the triangles DAC and CBE. These two triangles are thus also equicomplementable. It will hence be concluded that their bases AD and BE are also equal. Were this not the case and if it were assumed that, say $AD' = BE$, then by the well-known method of Euclid (cf. p. 62) it could follow that both triangles $AD'C$ and BEC are equicomplementable. Then, however, the triangles ADC and $AD'C$

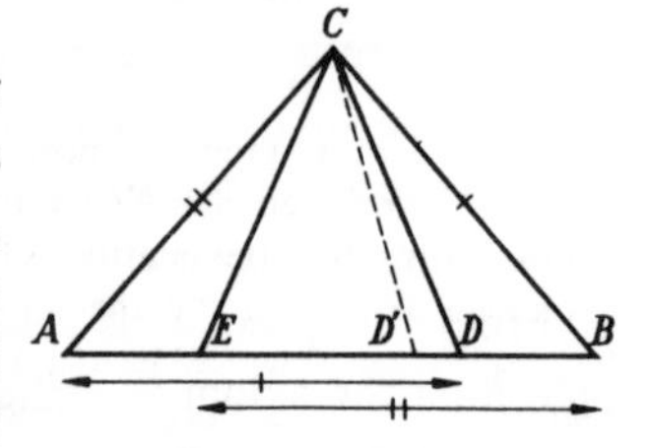

could also be equicomplementable. This would contradict the previously deduced theorem from the axiom of embedment. The equality of the segments AD and BE leads immediately to the assertion under consideration.

If in a triangle ABC the two sides AC and BC are equal then so are the angles opposite them.

For the proof, assume to the contrary that the angle $\measuredangle CAB$ is greater than the angle $\measuredangle CBA$. Determine then on the line AB the points A' and B' such that

$$\measuredangle CA'B = \measuredangle CBA \text{ and } \measuredangle CB'A = \measuredangle CAB.$$

Therefore, by the previously proved theorem

$$CA' = CB \text{ and } CB' = CA,$$

and hence follows, with the use of the hypothesis, that

$$(1^*) \qquad CA' = CB'.$$

Applying the exterior angle theorem to the triangles ACA' and BCB' one obtains the equations

$$\measuredangle ACA' = \measuredangle \mathrm{CAB} - \measuredangle CA'B$$

and

$$\measuredangle BCB' = \measuredangle CB'A - \measuredangle CBA.$$

Hence

$$(2^*) \qquad \measuredangle ACA' = \measuredangle BCB'.$$

Formulas (1*) and (2*) together with the hypothesis show that the triangles ACA' and BCB' are congruent in the narrower sense. But then, in particular, it would follow that

$$\measuredangle AA'C = \measuredangle BB'C.$$

This consequence is an absurdity, since both of these angles, namely, the interior angle and the nonadjacent exterior angle, are in the triangle $A'B'C$.

The stated theorem is thus proved, and it is seen at the same time that the congruence theorems in the broader sense are necessarily a consequence of the congruence axiom in the narrower form III, 5*, provided the aid of the above intuitive axiom of embedment concerning equidecomposability is enlisted."

From this proof it is clear at the same time that for the present application the axiom of embedment can be replaced by the following simpler axiom:

If in a triangle ABC the point D lies on the side AB between A and B then the triangles ABC and ADC are not equicomplementable.[1]

[1] A line of reasoning analogous to the argument for this result was used by me in the article "Über die Verwendung der Polygoninhalte an Stelle eines Spiegelungsaxioms in der Axiomatik der Planimetrie," *Elem. d. Math.*, Vol. 8 (1953), pp. 102-107. During the preparation of this work, the previous version of Appendix II was not available to me. As a result, an appropriate reference in the cited article was not given.

INDEX